Autorenkollektiv

Plasma Chemistry IV

Kinetics and Catalysis in Plasma Chemistry

M. Venugopalan and S. Veprek

High Pressure Plasmas and their Application to Ceramic Technology

P. Fauchais, E. Bourdin, J. F. Coudert and R. McPherson

AKADEMIE-VERLAG · BERLIN

Autorenkollektiv

Plasma Chemistry IV

Autorenkollektiv

PLASMA CHEMISTRY IV

With 122 Figures and 26 Tables

AKADEMIE-VERLAG · BERLIN
1983

Die Originalausgabe erscheint im Springer-Verlag Berlin · Heidelberg · New York

Vertrieb ausschließlich für die DDR und die sozialistischen Länder

Erschienen im Akademie-Verlag, DDR-1086 Berlin, Leipziger Straße 3—4
Lizenznummer: 202 · 100/528/83
Gesamtherstellung: VEB Druckerei „Thomas Müntzer", 5820 Bad Langensalza
Umschlaggestaltung: Karl Salzbrunn
Bestellnummer: 763 240 0 (6768) · LSV 1205
Printed in GDR
DDR 76,— M

Table of Contents

Kinetics and Catalysis in Plasma Chemistry

Mundiyath Venugopalan[1] and Stanislav Vepřek[2]

1 Chemistry Department, Western Illinois University, Macomb, Illinois 61 455, USA
2 Institute of Inorganic Chemistry, University of Zürich, 8057 Zürich, Switzerland

Table of Contents

Topics in Current Chemistry, Vol. 107
Managing Editor: Dr. F. L. Boschke

1 Introduction

For nearly fifty years the kinetics of reactions in electrical discharge plasmas has been intensively studied. These studies were mainly concerned with changes in rates of reactions due to variation of gas pressure, discharge current and reactant composition, and were chiefly applied to learn about the reaction mechanism. A typical and serious drawback [1] of homogeneous plasma reactions is their low energy yields and a pure selectivity. Since excitation in the plasma is not selective, a significant part of the electrical energy is consumed by the excitation of non-reacting low energy states. Preferential excitation of high energy electronic states will increase the chemical yield but not energy yield.

It has been pointed out [2,3] that catalytic processes are the only realistic way of increasing the reactor yield. Indeed, under plasma conditions it is experimentally found [4] that the catalyst in a heterogeneous reaction shifts the steady-state concentrations of the product. There is no analogue for this in the field of thermal catalysis where the catalyst merely diminishes the relaxation time, without affecting the equilibrium concentration of the product.

Another experimental problem in studying plasma reactions is that they may also be catalyzed by the surface of the reaction vessel and by electrode materials. The mechanisms of surface-catalyzed reactions must involve a sequence of steps:
1) adsorption of the reacting species on the surface,
2) reaction on the surface, and
3) desorption of the products.

Clearly, if one wants to study a homogeneous plasma reaction, it is important to find conditions under which the surface-catalyzed reactions are negligible. This is the case with the plasma synthesis of ozone which is now receiving considerable attention [5,6].

Because of the complex nature of plasma reactions in this article we shall first examine the kinetic and mechanistic aspects of some reactions. The kinetics of the dissociation of a number of molecules in plasmas has been reviewed recently [7]. Therefore, detailed studies of the environmentally and commercially important reactions of nitrogen with oxygen and hydrogen are presented here. These are the reactions which have been studied extensively in connection with heterogeneous plasma catalysis. The kinetics of other relevant plasma reactions are discussed briefly in the later sections where homogeneous and heterogeneous catalytic effects are discussed. The section on homogeneous catalysis includes a survey of the effects of added gases in plasma decomposition reactions. Following the classification originally given by Venugopalan [8] for a plasma-chemical process, heterogeneous catalytic effects have been discussed under three categories:
1) effects occurring in the plasma itself,
2) effects originating in the spatial afterglow region, and
3) effects taking place in the quenching or cold trap stage.

In a final section the effects of plasma-treatment of surfaces and materials including catalysts are described.

2 Kinetics of Some Plasma Reactions

2.1 The Nitrogen-Oxygen System

The reaction of nitrogen and oxygen in a plasma produces significant amounts of NO in the gas phase [9]. At gas pressures greater than 10 Torr some NO_2 has also been reported [10–12], but at lower pressures NO_2 is formed usually when the plasma gas is quenched to temperatures below 175 K [13]. Even at the higher pressures the NO_2 yield decreases rapidly with increasing residence time in the discharge and becomes almost negligible at $t > 30$ ms [10]. The principle of local thermodynamic equilibrium limits the concentration of nitric oxide in the plasma in air to 3–5 vol % at T = 4000 to 6000 K and p = 50–760 torr [14], but concentrations as high as 12–15 vol % have been reported [11] using high-current pulsed discharges of less than 10^{-7} s duration.

Vasil'ev and coworkers [15] measured the volume concentration of NO in a discharge, using stainless steel electrodes, as a function of reaction time at constant power and found that the reaction

$$N_2 + O_2 \underset{k_2}{\overset{k_1}{\rightleftharpoons}} 2NO \tag{1}$$

is representable by a "first" order which is indicative of a direct electron impact process and/or a vibrationally excited reaction channel. The reaction time was deduced from the ratio of reaction volume to the volume flow rate of the gases. Since the rate of formation of NO was found to be dependent on the plasma power U, they suggested that

$$[NO] = (k_1/k_2)\,[1 - \exp(-k_2U/v)] \tag{2}$$

where U/v is termed the specific energy Wh dm^{-3} and k_1 and k_2 are the rate constants ($dm^3\ Wh^{-1}$) for the forward and reverse reactions, respectively. Equation (2) can be deduced from the fundamental principles of chemical kinetics and the fact that the reaction rate is proportional to the plasma power if a zero order of formation of NO is postulated, i.e. changes in initial concentrations of N_2 and O_2 are ignored, which is justifiable if the product concentrations are relatively small. It follows from eq. (2) that as $U/v \to 0$

$$[NO] = k_1(U/v) \tag{3}$$

and as $U/v \to \infty$

$$[NO] = [NO]_{ss} = k_1/k_2 \tag{4}$$

Typical kinetic curves [4,16] shown in Fig. 1 also suggest that with large specific energies, i.e. long reaction times, a steady state is reached in which the concentration of the product is independent of the specific energy. Thus, with nitric oxide the

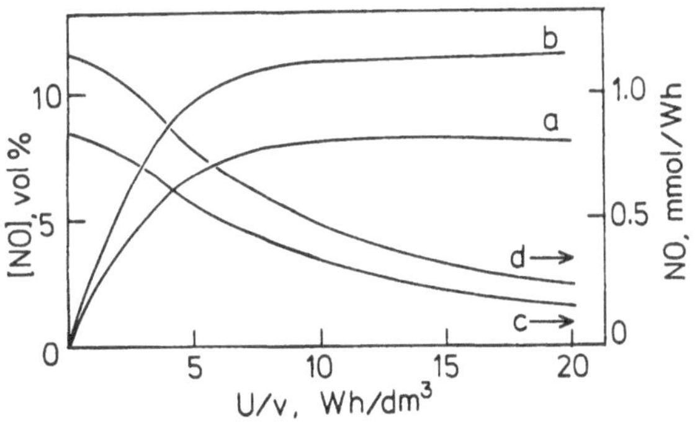

Fig. 1. [NO] versus U/v curves for (a) air and (b) stoichiometric $N_2 + O_2$ mixture. (c) and (d) curves represent the energy efficiency of NO formation in mmol/Wh for the same mixtures [4)]

steady state concentration reaches 11–12 vol % under the given conditions for a stoichiometric gas mixture. This steady-state concentration, at which rates of formation and the decomposition of NO are equal, exceeds the thermodynamic equilibrium concentration which is calculated to be 6.5 vol %.

The steady state concentration depends on many factors. Figure 2 shows that $[NO]_{ss}$ increased directly with the product *ip* (*i* is the current in amperes and *p* is the gas pressure in torr) up to $\simeq 20$ A torr, after which it either reached a limiting value (at $p < 100$ torr) or passed through a maximum value (at $p > 100$ torr) [17)]. The divergence of the curves at large *ip* values apparently is the result of a decrease in the reduced field E/p with an increase in the gas pressure [18,19)]. Since E/p serves as a measure of the electronic temperature T_e in the plasma plots of E/p and $[NO]_{ss}$ against *ip*V for air are compared in Fig. 3. Here V is the potential drop across the discharge gap. Note that the relation between E/p and *ip*V is represented by a curve common to all the pressures, describing the fall of the relative potential gradient with increase in *ip*V. Furthermore, the same figure shows that at all pressures the

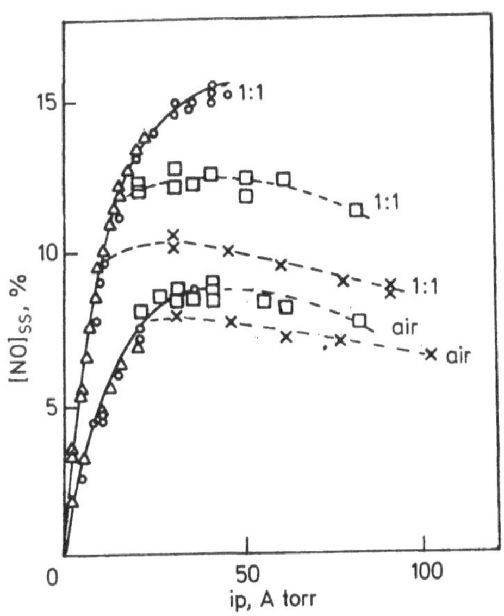

Fig. 2. Variation of $[NO]_{ss}$ with ip (A torr) for air (bottom) and stoichiometrix $N_2 + O_2$ mixture (to p). △, ○, □ and × represent data for 50, 100, 200 and 300 torr, respectively [17)]

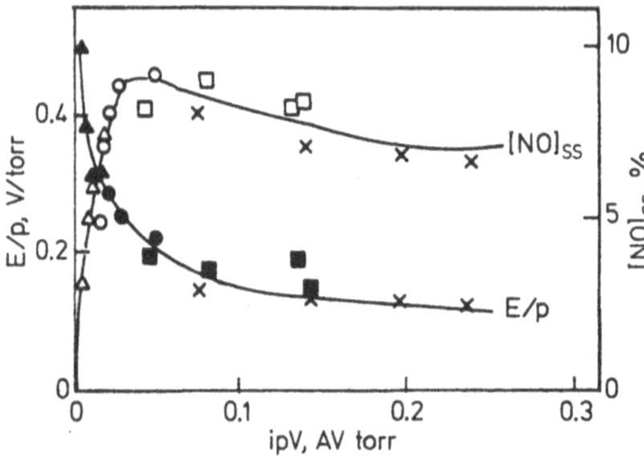

Fig. 3. Variation of E/p and $[NO]_{ss}$ with ipV at the different pressures given in Fig. 2 [19]

dependence of the limiting concentrations on *ip*V is approximately described by one curve, which passes through a maximum.

The relations between k_1 and k_2, on the one hand, and *ip*V, on the other, are quite different (Fig. 4). Clearly, the conditions for the attainment of high steady-state concentrations, i.e. the highest values of the equilibrium constant do not coincide with the conditions for the highest rate of formation of nitric oxide, i.e. highest k_1 which corresponds to a pressure of 300 torr and a current of 100 mA. Under the latter conditions the yields of NO are highest at the beginning of the kinetic curve, i.e. low U/v.

Emel'yanov and coworkers [20] have determined the gas temperature T_g from the molecular spectrum of a discharge in air and found that the change in $[NO]_{ss}$ is almost entirely related to the change of T_g alone in the temperature range of

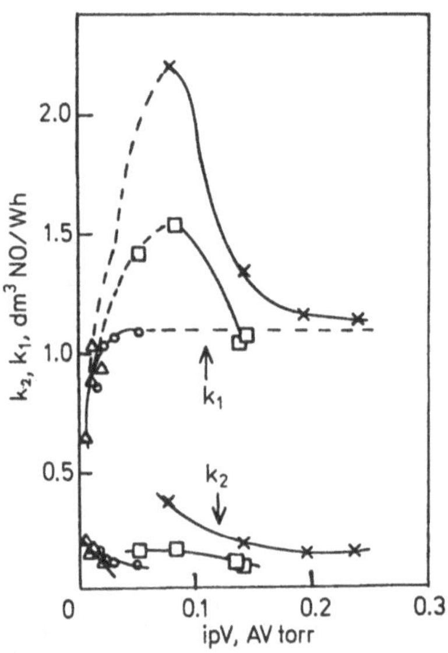

Fig. 4. Variation of k_1 (top) and k_2 (bottom) with ipV at the different pressures given in Fig. 2 [19]

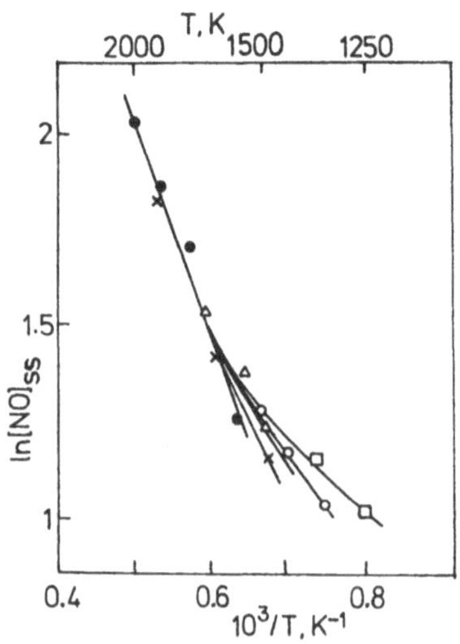

Fig. 5. Variation of ln $[NO]_{ss}$ with $1/T_g$ at different pressures. □, ○, △, ×, ● represent 15, 20, 30, 50 and 100 torr, respectively [20]

1550–2000 K, regardless of the gas pressure or density (Fig. 5). By including this temperature dependence in calculating the kinetic constants they found that k_1 is independent of T_g, whereas k_2 has a negative temperature coefficient [21]. The observed increase in $[NO]_{ss}$ with rise in temperature of the gas in the reaction zone was explained by the accompanying decrease in k_2 while k_1 remained unchanged. It was also demonstrated [22] that the role of thermal reactions involving the formation and decomposition of NO is insignificant in the experimental range of discharge powers and gas temperature (<2000 K) investigated.

Table 1 shows that $[NO]_{ss}$ also depends on gas composition and the diameter d of the reactor [17]. With regard to gas composition a stoichiometric mixture of N_2 and O_2 rather than "reciprocal air" ($N_2/O_2 = 0.25$) is desirable. The narrower the discharge tube the greater is the steady state concentration. Indeed, an investigation [23] has shown that $[NO]_{ss}$ increases linearly with 1/d, which indicated the increasing importance of the heterogeneous component of its formation.

Table 1. Steady-state concentration of nitric oxide as a function of gas composition and pressure in different reactors [17]

p	N_2/O_2	4		1		0.25	
torr	d (mm)	3	20	3	20	3	20
50		7.2	5.0	14.0	8.0	10.3	6.8
100		9.3	5.5	15.5	8.1	12.1	6.8
200		9.0	5.5	12.8	7.9	9.8	6.1
300		8.3	5.0	11.0	7.2	8.8	5.5

2.2 The Nitrogen-Hydrogen System

The reaction of nitrogen and hydrogen in a plasma gives ammonia, the decomposition of which produces hydrazine.

2.2.1 Synthesis of Ammonia

In a dc discharge in $N_2 + H_2$ mixture Brewer and Westhaver [24] found that NH_3 was formed only in the luminous discharge regions, and predominantly in the negative glow. Figure 6 illustrates the dependence of the degree of conversion of a stoichiometric $N_2 + H_2$ mixture into ammonia on specific energy at 5 torr under flow conditions in a water-cooled quartz reactor with electrodes 5 cm apart and discharge currents of 30 and 60 mA [25]. With such short interelectrode separations probably only the negative glow and the dark spaces adjacent to each electrode are present [26]. The figure also shows the electron temperatures determined by a double probe method. The probes were made of W wire of 0.2 mm diameter and were 2 mm apart. A steady-state degree of conversion (0.66 vol %) is reached at values of electron temperatures ($\simeq$1.8 eV) which are independent of specific energy (U/v > 15 Wh dm^{-3}). Also, the concentration of electrons n_e, as determined by the double probe

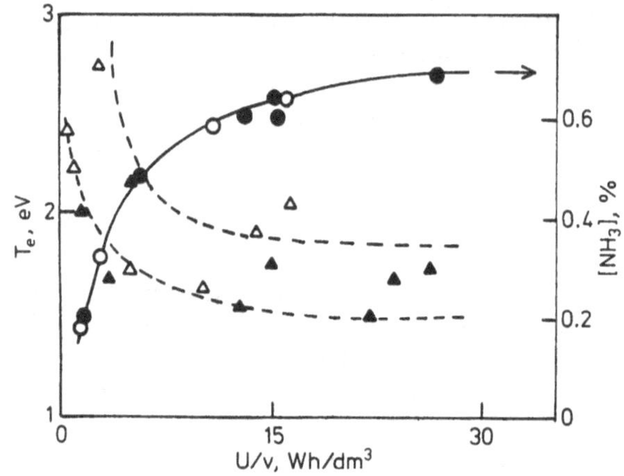

Fig. 6. Dependence of degree of conversion (circles) of a stoichiometric $N_2 + H_2$ mixture into NH_3 on specific energy and electron temperature (triangles) for $n_e = 1.2 \times 10^9$ and $2.5 \times 10^9\ cm^{-3}$ at discharge currents of 30 (open points) and 60 mA (closed points), respectively [25]

method, remained almost constant over the given specific energy range, but was higher the larger the discharge current. Under these conditions the *rate of synthesis of ammonia* is described by the equation

$$\begin{aligned} d[NH_3]/dt &= k_1'[N_2]\,[H_2] - k_2[NH_3] \\ &= k_1 - k_2[NH_3] \end{aligned} \tag{5}$$

in which $k_1 = k_1'[N_2]\,[H_2] = 0.22\ dm^3\ Wh^{-1}$

and $k_2 = 0.34\ dm^3\ Wh^{-1}$

are the rate constants for formation and decomposition, respectively, of NH_3.

Since a glow discharge in $N_2 + H_2$ mixtures exhibits emission lines due to NH [27] the ammonia is formed by the addition of a molecule of H_2 to NH:

$$NH + H_2 \rightarrow NH_3^* \tag{6}$$

The NH_3^* molecules formed by reaction (6) have excess energy which must be removed for their stabilization, perhaps by quenching of the reaction mixture to low temperatures [27]. Otherwise, they will be destroyed by the reaction with H atoms:

$$NH_3 + H \rightarrow NH_2 + H_2 \tag{7}$$

In fact emission lines due to NH_2 have been reported [27]. The values of k_1 and k_2 suggest that reaction (7) was preponderant over reaction (6) under the given conditions. From the synthetic point of view a pressure of 100 torr appeared to be optimum and a gas composition of $N_2:H_2 = 5:1$ was reported somewhat unexpectedly to result in the highest NH_3 yield [28].

2.2.2 Decomposition of Ammonia

In a dc discharge in ammonia significant amounts of hydrazine were formed only in the positive column [29,30]. This is to be compared with the observation [24] that the concentration of NH_3 was a maximum in the negative glow and evenly distributed in low concentration in the positive column of a discharge in $N_2 + H_2$ mixtures. Such selective distribution of reaction can be taken advantage of in avoiding hydrazine formation during ammonia synthesis.

In closed systems [31] pressure changes during the discharge were found to be associated with a decrease in the average energy of electrons (Fig. 7). The electron energies were measured by the double floating probe method using W probes 0.05 mm in diameter. However, the decrease of the average energy could not be attributed solely to the increase of gas pressure because the decrease of T_e should be slower on the basis of the experimental data (curve d, Fig. 7) corresponding to the initial instants during discharge and to different initial pressures. It was explained [32,33] by a change in the composition of the gas on formation of negative ions as a result of the attachment of electrons to ammonia molecules. The decomposition of ammonia molecules presumably decreased the concentration of negative ions and lowered the diffusion coefficient of electrons and their average energy. Similar results were reported [31] for a glow discharge at 30 mA, in which T_e and E/p decreased continuously with the duration of discharge.

Simonyan et al. [34] have determined the energy distribution function for *electrons in a glow discharge* in NH_3 under flow conditions using a probe made of Pt wire 0.1 mm in diameter. In Fig. 8 the normalized energy distribution functions for electrons in the ammonia plasma are compared with the Maxwell distribution functions for the given average energy. When the gas flow rate was increased from 4 to 10 $dm^3\ h^{-1}$, a slight alteration of the distribution function was observed

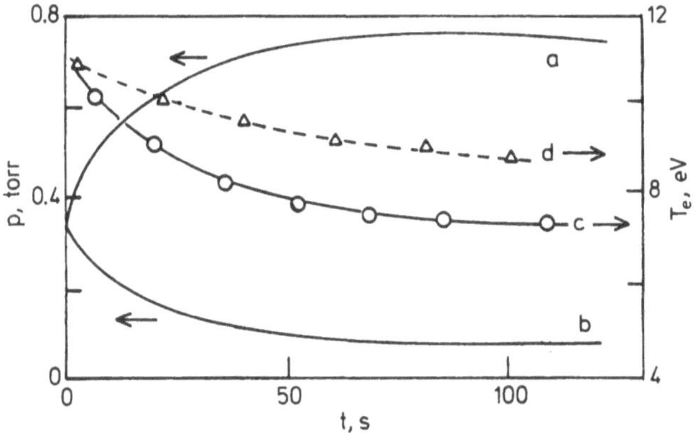

Fig. 7. Time variation of (a) the overall pressure p_{total}, (b) the partial pressure of ammonia p_{NH_3} and (c) the average energy of electrons T_e during an 80 W discharge at 39 MHz. Curve (d) represents the variation of the average energy associated solely with the increase of the overall pressure [31]

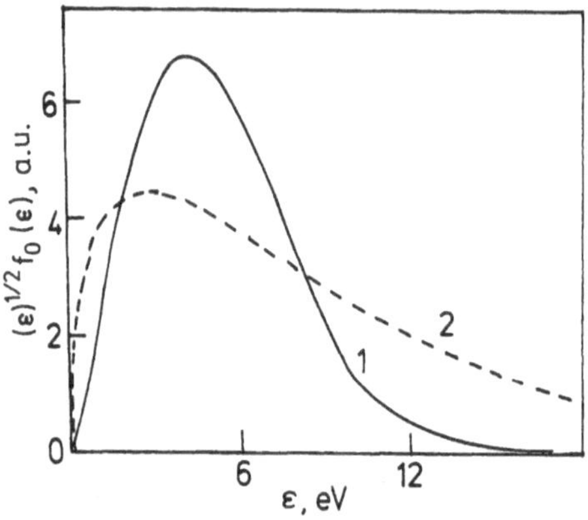

Fig. 8. The electron distribution function with respect to energies in a glow discharge in ammonia at 3 torr and 5 mA.
(1) experimental; (2) Maxwellian [34]

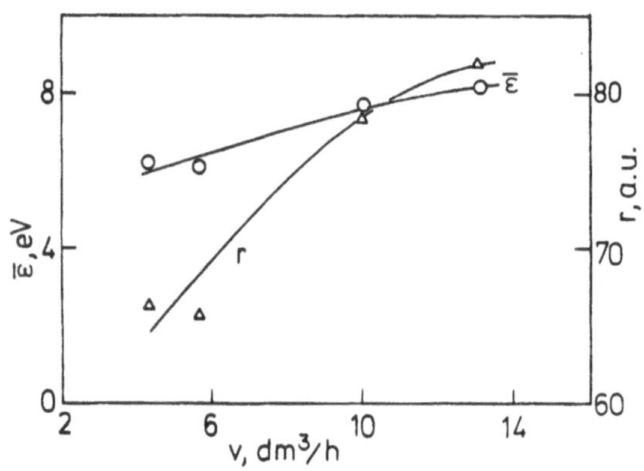

Fig. 9. Calculated rate constants r for the excitation of NH_3 by electrons and the average electron/energy for different gas flow rates v [34]

in the range of energies up to 8 eV, while the average energy of the electrons remained 6.05 eV. At high gas flow rates, an inflection or an additional maximum appeared on the distribution curves; the fraction of slow electrons decreased (in the range 0.5–5.5 eV) with a consequent increase of the average energy. The distribution functions were used to calculate the rate constants shown in Fig. 9 for the excitation of NH_3 by electrons:

$$NH_3 + e \xrightarrow{r} NH_3^* \tag{8}$$

Evidently with an increase of the flow rate both the average electron energy $\bar{\varepsilon}$ and the rate constant r for the excitation of NH_3 increased. Since n_e ($\simeq 3 \times 10^9$ cm^{-3}) under the experimental conditions varied only within the limits of experimental error, the decomposition rate constant k_1 ($= rn_e$) must depend on the specific energy U/v of the discharge.

A study [35] of the plasma decomposition of NH_3 in a 50 Hz glow discharge showed that the relation between the overall degree of conversion of ammonia and the

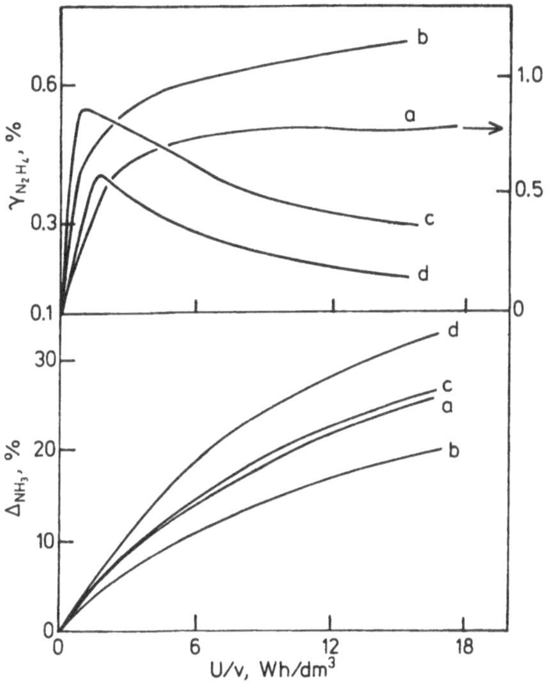

Fig. 10. Variation of the overall degree of decomposition Δ% of ammonia (bottom) and its conversion γ% into hydrazine (top) with specific energy at pressures of (a) 5, (b) 10, (c) 25 and (d) 50 torr. Data at a current of 40 mA for a reactor of 3 mm diameter except at p = 5 torr for which a reactor of 12 mm diameter was used [35)]

specific energy are similar over a wide range of currents (0–80 mA) and pressures (0–50 torr), while the yields of N_2H_4 can differ appreciably (Fig. 10). In general the yield of hydrazine decreased with increase of pressure and this was attributed to its decomposition in the discharge. However, the results also suggested that hydrazine is not an intermediate in the decomposition of NH_3 to the elements in the electric discharge and the synthesis of hydrazine and the formation of N_2 and H_2 proceed via parallel reactions involving NH and NH_3 or two NH_2 radicals. A kinetic calculation showed that the decomposition of ammonia is a first order process which indicated that the primary step is the excitation of the ammonia molecule on collision with an electron (reaction 8).

In another study [36)] at low pressure ($p < 2$ torr) the yield of hydrazine was found to decrease with residence time in a discharge which emitted both H and NH_2 lines (Fig. 11) suggesting that the combination of NH_2 radicals leads to either N_2H_4 or NH and NH_3. Photoelectric measurements [37)] of NH_2 absorption as a function of time following an rf discharge pulse through ammonia showed that the kinetics of NH_2 disappearance to form hydrazine is second order with a rate constant of $\simeq 2.3 \times 10^9$ dm^3 $mole^{-1}$ s^{-1}. From the synthetic point of view the yield could thus be increased by reducing the residence time of the hydrazine in the discharge in agreement with the general premise of hydrazine degradation in the discharge [38)]. When the flow rate is high the formation of N_2H_4 may be occurring primarily in the cold trap so that there is no opportunity for the discharge to affect the decomposition. An alternative explanation [39)] is based on the reaction

$$N_2H_4 + H = NH_3 + H_2 \tag{9}$$

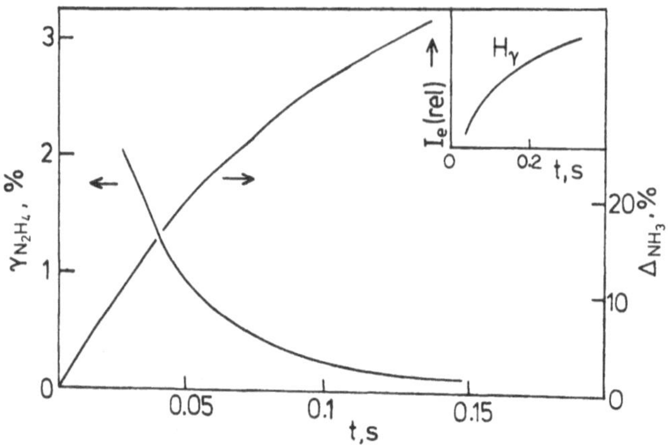

Fig. 11. Percent decomposition (Δ) of NH_3 and N_2H_4 yield (γ) as a function of residence time (t) of NH_3 in the discharge. Inset shows the relative intensity of H_γ (4341 Å) emission at 49.5 mA and 1.7 torr [36)]

and is supported by the increase in H atom emission intensities with residence time (inset Fig. 11).

In yet another study [40)] in the pressure range 5–40 torr, but using a capacitively coupled rf discharge, the decomposition of NH_3 followed an apparent zero order kinetics which was interpreted on the basis of a kinetic mechanism involving vibro-rotationally excited molecules. Figure 12 shows the degree of conversion of NH_3 as a function of the mean residence time within a tubular reactor. Simultaneous spectroscopic measurements at different positions along the reactor axis showed

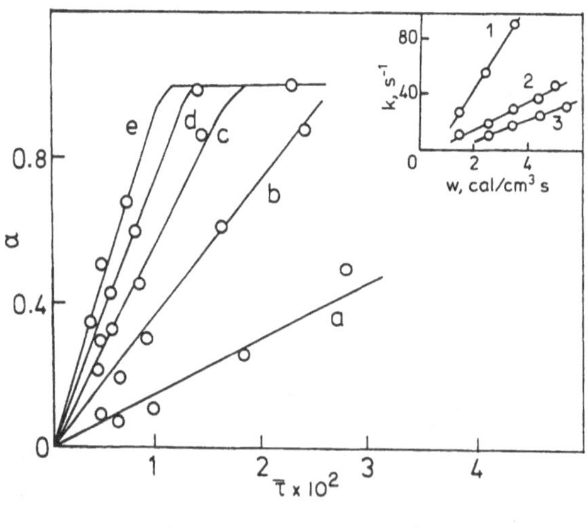

Fig. 12. Power dependence of the degree of conversion α of NH_3 (p = 20 torr) as a function of the mean reidence time $\bar{\tau}$. Curves a, b, c, d and e for power densities of 1.5, 2.5, 3.5, 4.5 and 5 cal/cm³ s, respectively. Inset shows rate constant k (s^{-1}) of the primary step of NH_3 decomposition as a function of power density at pressures of (1) 5, (2) 20 and (3) 40 torr [40)]

a remarkable disequilibrium existing between the various degrees of freedom: the vibrational temperatures, which were determined from the relative values of the band intensity of the sequences $\Delta v = +2, +1, -2$ of the system $N_2(C^3\Pi_u - B^3\Pi_g)$, had a mean value of $T_v = 3500$ K; rotational temperatures, which were evaluated from the relative values of the rotational line intensities of the vibrational band (0, 0) of the $NH(c^1\Pi - a^1\Delta)$ system and (0, 2) of the $N_2(C^3\Pi_u - B^3\Pi_g)$ system had a constant mean value of $T_R = 2000$ K. Both temperatures were independent of the axial position and power density.

The variation of the intensity of NH along the axis was found to be typical for an intermediate species while those for N_2 and H_2 were typical of species whose concentration increased with residence time (Fig. 13). Values of the rate constants, derived from the slope of the plots of Fig. 12 are shown in the inset as a function of the power density; these values are several orders of magnitude higher than those usually reported for thermal decomposition of ammonia. Since emission due to NH_2 could not be assigned and measured with certainty, it was not possible to discriminate between the mechanisms:

$$NH_3^* \rightarrow NH + H_2 \tag{10}$$

$$NH_3^* \rightarrow NH_2 + H \tag{11}$$

Nevertheless it was shown that a reaction scheme including either of these steps would lead to a zero order kinetics depending on the value of the rate constant for dissociation of hydrogen. Possibly both mechanisms were operative under the experimental conditions.

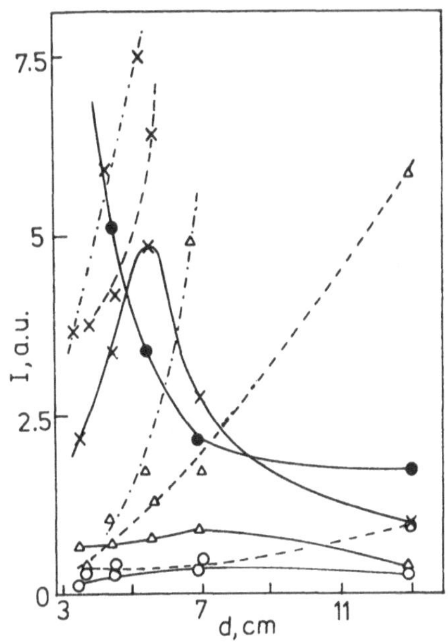

Fig. 13. Intensity I of N_2 $(C^3\Pi_u - B^3\Pi_g)$ (dashed lines), $NH(A^3\Pi - X^3\Sigma^-)$ (solid lines) and $H_2(G^1\Sigma_g^+ - B^1\Sigma_u^+)$ (broken line) in arbitrary units along the reactor axis d at various power densities. ○, △, ×, ● represent 0.44, 0.75, 1.3 and 2.13 cal/cm³ s, respectively [40]

Other spectroscopic studies [41–44] of the glow discharge in ammonia have confirmed that the reaction

$$NH(^1\Delta) + NH_3 \rightarrow N_2H_4 \quad (12)$$

predominates in the *plasma synthesis of hydrazine*. Since decomposition of ammonia molecule in the discharge to NH and H_2 involves the formation of only one $NH(^1\Delta)$ for every three $NH(^3\Sigma)$ which are inactive in the synthesis of hydrazine [45], the hydrazine yields will be low as found experimentally. In an ac discharge in ammonia the relative intensities in the rotational structure of the electronic transition $NH(^3\Pi - {}^3\Sigma)$ gave temperatures in the range 600–1400 K (evidently close to the temperature of the gas) which fell only slightly with increase in specific energy [43]. Considerable changes in the yield of hydrazine (Fig. 10) with variation in the specific energy while the gas temperature changed very little indicated that thermal activation plays a negligible part in the synthesis of hydrazine in the discharge. Furthermore, a calculation [43] shows that the thermodynamic equilibrium concentration of hydrazine at 600 K is $\simeq 2 \times 10^{-7}$ vol %, whereas the reported concentrations are $\simeq 0.5$ vol %.

In the same studies [44] a determination of the vibrational temperature from the relative intensities of bands in the second positive system of N_2 gave values in the range 2200–5000 K which did not change with increase in specific energy. Since the rotational temperature, probably close to the gas temperature, was almost 90% lower than the vibrational temperature, the plasma was non-isothermal under the experimental conditions which favored hydrazine formation. The non-equilibrium population of the highest vibrational levels of the nitrogen molecule has been adduced [44] as indirect confirmation of the theory of energy catalysis (see Sect. 3).

3 Homogeneous Catalysis

In this section we consider the enhancement of chemical conversion and product selectivity due to a foreign gaseous substance added to the plasma. Such a catalytic or sensitizing effect under homogeneous conditions has been attributed to energy transfer or charge transfer from highly excited and ionic species of the additive gases [46]. Typical additives which have been studied include the noble gases, H_2, N_2, H_2O, CO_2 and Hg. An unequivocal relation between the degree of catalysis and ionization potential has, however, been observed only in the case of the noble gases.

3.1 Noble Gas Plasmas

In an attempt to elucidate the fundamental activating process of homogeneous catalysis in plasmas Rubtsova and Eremin [47,48] studied negative corona discharges in xenon whose electrical characteristics are very sensitive to changes in gas composition [49,50]. A dc corona discharge was produced in a static system at a negatively charged pointed Pt electrode 0.2 mm in diameter separated by a 6 mm gap from a flat nickel-plated anode 20 mm in diameter. The threshold voltages in mixtures of xenon with O_2, CO_2, N_2, and air are shown in Fig. 14 as functions of concentrations

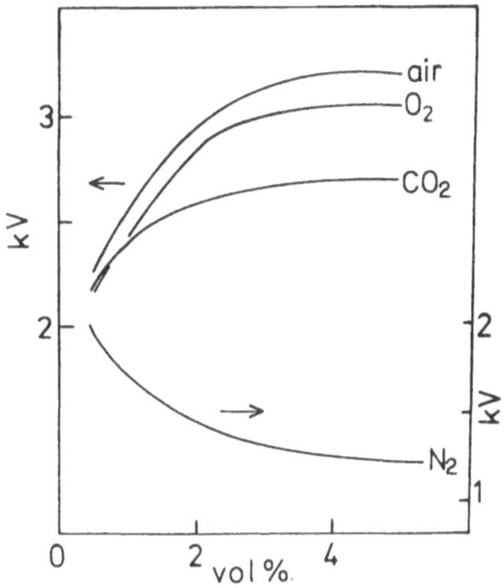

Fig. 14. Threshold (initial) voltage (kV) of the negative corona discharge in a mixture of xenon with added O_2, CO_2, N_2 and air (vol %) at 500 torr [47,48]

of O_2, CO_2, N_2 and air, respectively. Air, like O_2 and CO_2, increased the threshold voltage, whereas N_2 lowered it over the pressure range 250–570 torr.

Since the *corona discharge* was not observed in spectroscopically pure xenon before direct breakdown of the discharge gap occurred, it was concluded [47] that the presence of molecular gases is essential to the formation of a negative corona discharge in xenon. Some gases (O_2, CO_2, air) form negative space charges which can carry the current in the outer region of the corona. Under these conditions the threshold voltage increases with the magnitude of the space charge and with decreasing mobility of the constituent ions. On the other hand, in the presence of N_2 negative space charges are not formed, and the current is carried to the anode (outside the immediate vicinity of the pointed electrode) by the slow electrons formed by inelastic collisions with the particles of the molecular gases. That the effect of CO_2 on the threshold voltage of the corona is weaker than that of O_2 is unexpected, since CO_2 is heavier and has a higher electron affinity.

3.2 Plasmas in Diatomic Gases

3.2.1 Hydrogen Plasmas

Plasma dissociation of hydrogen has been studied extensively [7]. However, little is known about the catalytic effects on the dissociation under homogeneous conditions. The effects of oxygen and water vapor have been discussed [51,52], but the mechanism of the process is not clear. It was suggested that ionization potentials are important factors in-so-far as the effect of H_2O is concerned [52].

Zaitsev et al. [53] determined the distribution function of the energies of electrons in a hydrogen discharge (p = 0.258 torr, i = 15 mA) and found that the presence of water vapor and chlorine lead to an increase in the fraction of high-energy electrons (a second maximum at 10–12 eV of the distribution function). The increase was

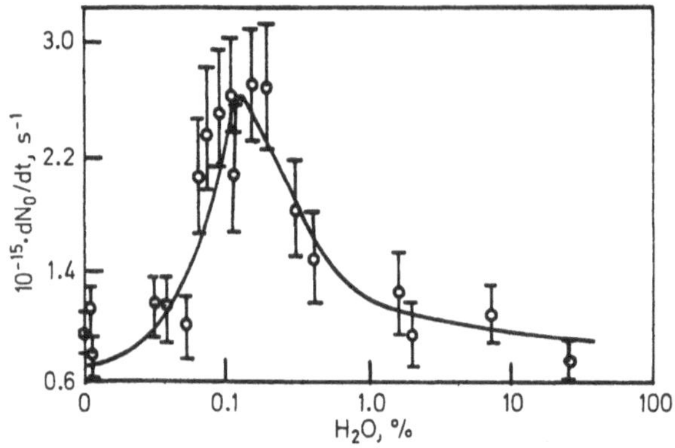

Fig. 15. Variation of the rate of dissociation of hydrogen with water vapor concentration. The curve represents calculated values; circles are experimental values [53]

attributed to a change in the character of the diffusion of electrons in the presence of negative ions. The experimental distributions and the concentrations of the electrons were used to calculate [54] the rate of dissociation of the hydrogen at various concentrations of water vapor. The experimental rates of dissociation shown in Fig. 15 were measured by the method of selective absorption of atoms in a similar discharge. The rate of dissociation is maximum at a H_2O concentration of $\simeq 0.15\%$ and is approximately three times the rate of dissociation of pure H_2. Clearly, one of the reasons for the effect of additions of water on the yield of H atoms in a glow discharge is the change in the electron energy distribution function.

The kinetics of the loss of negative ions [55] showed that increase in the additive content to 10% is accompanied by change in the recombination coefficient of H atoms on the wall and increase in the degree of dissociation of the molecules.

3.2.2 Oxygen Plasmas

In the plasma synthesis of ozone from O_2 + Ar mixture [56] the degree of conversion of O_2 to O_3 was independent of the composition of the mixture indicating that Ar does not play any active part in the reaction. However, the concentration of ozone formed from oxygen in a silent electric discharge was increased by the addition of N_2 [57], the maximum concentration being obtained from a N_2:O_2 mixture in the ratio 1:4 (Fig. 16). The explanation [58] given for the activating effect of N_2 is that

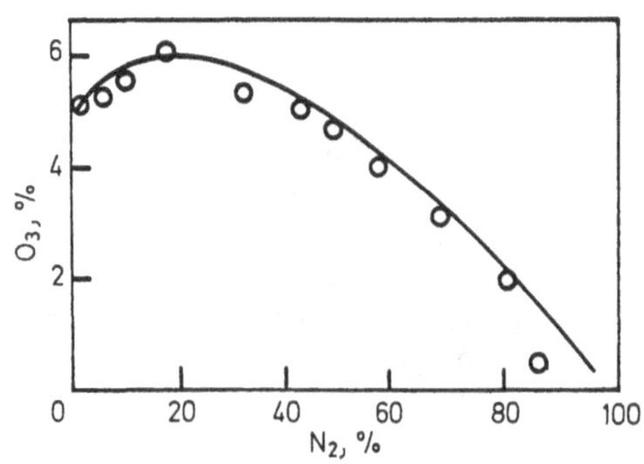

Fig. 16. Dependence of concentration of ozone on composition of mixture. The curve is calculated using the kinetic scheme given in Ref. [59]

it can function as an energy catalyst. The dependence of the intensity of the nitrogen bands on the composition of the mixture suggested the role of excited N_2 in the synthesis of ozone [59]. Certainly NO, NO_2 and N_2O_5 must be formed in these discharges. In pulsed discharges through air at atmospheric pressure and low flow rates the effect of NO_2 poisoning of the ozone production was found to be important [60]. However, in an ozonizer (barrier) discharge under flow conditions the concentration of N_2O_5 corresponding to the disappearance of O_3 was greater than the steady-state concentration of the lower oxides [61].

3.2.3. Halogen Plasmas

It is found that the dissociation of chlorine and bromine is greatly accelerated by the presence of He, Ar and Xe [62,63]. Since the increase in the dissociation rate was associated with a decrease of E/p it cannot be explained by an increase of the average electron energy. Consequently, it has been proposed that the noble gas atoms in their metastable states behave as energy catalysts. *Energy catalysis* is fairly effective only when the metastable atom X* gives up almost all its energy on collision with the molecule M_2:

$$X^* + M_2 \rightarrow X + M_2^* \rightarrow X + M + M$$

That is to say, energy catalysis must depend on the system under consideration.

3.3 Plasmas of Polyatomic Molecules

The plasma dissociation of ammonia was hardly affected by the presence of noble gases [62,63]. However, the presence of benzene has been shown [64] to greatly enhance the production of N_2 from ammonia. On the other hand, the total decomposition of benzene is markedly reduced by the presence of ammonia. In this system certainly there was some chemical reaction between benzene and ammonia. At the electron energies in the discharge the collision cross section for benzene ionization is higher than for ammonia ionization so that ammonia is unlikely to ionize in preference to benzene unless the ammonia is present in large excess.

In the high frequency decomposition of water vapor [65] Ar was found to be an inert diluent, but CO_2 produced a strong inhibitory effect on the production of hydrogen peroxide in a cold trap [66]. The inhibitory effect could, however, be minimized by hydrogen discharge cleaning suggesting that surface poisoning was also involved [67]. Brooks and Sambrook [68] have found that the addition of CO_2 to a microwave discharge in benzene resulted in a lowering of the reaction rate and a change in product distribution. A similar effect was reported [69] in the microwave decomposition of cyclohexane: in this case the presence of CO_2 had a more pronounced effect on the distribution of reaction products. The carbon dioxide did not react chemically in any of these systems. Further, the ionization potential of CO_2 is higher than that of benzene or cyclohexane.

In the dissociation of CO_2 [70] and in the oxidation of CO [71] in a silent discharge,

Ar and He behaved as inert diluents, while N_2 acted as an "energy catalyst". However, in pulse discharges of CO_2, N_2 was found to play the role of an inert diluent [72].

The addition of water vapor in the plasma reaction of H_2 and CO has a strong inhibitory effect on the production of methane [73,74], but He and Ar increased the conversion of CO to CH_4 [75]. The addition of He accelerated the decomposition of CH_4 to a much greater extent than the addition of H_2 and this was explained by the fact that He plays the role of an energy catalyst [76,77]. From a mass spectrometric analysis of ionic reactions in a glow discharge of methane Drost and Klotz [78] concluded that a relatively small catalytic effect is caused by added gases such as He, Ar, N_2, H_2 and H_2O, primarily due to the additional formation of primary ions such as CH_2^+ and CH_3^+ by dissociative charge transfer reactions. Such reactions can occur only if the ionization potential of the sensitizing atoms or molecules is higher than that of the molecules of the plasma gas. The addition of He to a $N_2 + CH_4$ plasma increased the intensity of $N_2H_4^+$ ions by nearly one order of magnitude [75]. Luk'yanov et al. [79] have shown that, depending on the composition of a $CH_4 + CO_2$ plasma, Ar has different effects on the yields of carbonyl and carboxy products, behaving in some cases as a catalyst and in others as an inert diluent.

3.4 Synthesis of Ammonia

The synthesis of NH_3 has been studied [80,81] in the presence of metallic vapor. The effect of Hg vapor was positive, while Zn and Cd gave negative effects (Fig. 17). An explanation of the positive effect was offered on the basis of the reactions of excited Hg* (4.8 eV).

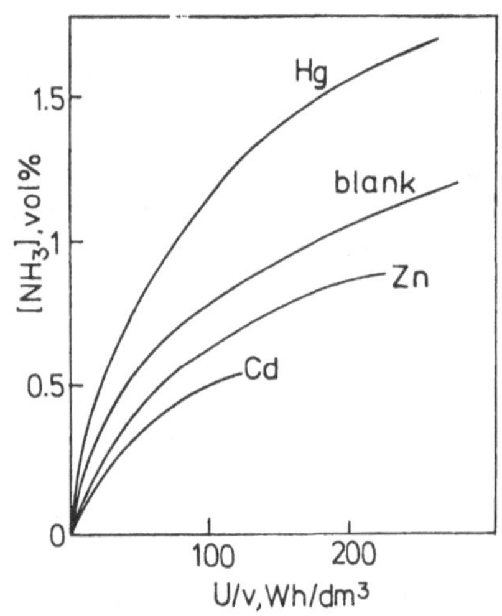

Fig. 17. Yield of NH_3 as a function of specific energy for different metallic vapor catalysts [78]

4 Heterogeneous Catalysis

In this Section we consider the enhancement of plasma-chemical conversion and product selectivity due to a foreign substance in the solid phase. The substance may be placed in the plasma, in the spatial afterglow, in a cold trap for collecting the products or in all of these regions. Typical solids which have been used include several transition metals and some of their oxides [4]. It appears that there is some connection between the degree of catalysis and the electron work functions of the metal catalysts. So far as conventional catalysis is concerned attempts to correlate the electronic factor with catalytic activity and chemisorption — the precursor of catalysis — has largely been unsuccessful (see Sect. 6).

4.1 Catalytic Effects Occurring in the Plasma Zone

Heterogeneous catalytic effects in the plasma zone may arise from impurities and coatings on electrodes, packing materials, grids placed in the inter-electrode space, and coated walls of the discharge tube. Such effects have been studied mainly in connection with the syntheses of O_3, H_2O_2, NO, NH_3, N_2H_4 and CH_4.

4.1.1. Effects Due to Materials on Electrodes

Besides supplying a contaminant to the plasma volume the electrodes may influence the reactions occurring therein by either accelerating or retarding their rates. A typical case under study for several decades has been the plasma synthesis of ozone.

4.1.1.1 Ozone Synthesis

As early as 1908 Warburg and Leithaeuser [82] reported that ozone concentrations and energy yields from an ozonizer containing a gold-plated metal electrode were lower than those obtained from an all-glass ozonizer. Later Starke [83] found that the concentration of ozone produced in an ozonizer with an Al electrode is only one-half that obtained in an all-glass ozonizer. More recently the use of a number of other metals for electrodes has been reported [84–86]. These include Fe, Mg, Zr, Hf, Va, Nb, Ta, Cr, Mo, Ti, stainless steel, Duralumin and anodized Duralumin. However, the steady-state concentrations of ozone were very similar for all the metals, at a level approximately one-half that achieved in an all-glass ozonizer. A possible explanation of this effect is that the oxide film formed on the metal surface during the plasma synthesis of ozone has no catalytic action and, therefore, roughly equal concentrations of ozone are produced on all the metal electrodes.

A metal ozonizer with a polyester rather than glass dielectric barrier gave higher ozone concentrations (40–50 *vs* 20 g m^{-3} air) and lower energy consumptions [87]. A kinetic treatment of the data has shown that in both types of ozonizers the rate of ozone destruction is of the same range, but the rate of production is higher in the ozonizer with a polyester layer. Naturally, the metal ozonizer has the additional advantage that is is less sensitive to shock and mechanical damages than glass ozonizers.

Salge and coworkers [88] have measured the local ozone distribution within the discharge gap using the relative absorption at $\lambda = 254$ nm. They observed an increase of the local ozone generation towards the glass dielectric. The increase was dependent on the voltage steepness applied to the ozonizer: smaller values reduced the effect and at sine waves the ozone concentration at the dielectric was remarkably decreased. It has been pointed out [89] that the beneficial effect of the short voltage pulse is a reduction in the ozone decomposition which otherwise occurs via electron attachment and a rise of the temperature.

4.1.1.2 Hydrogen Peroxide Synthesis

Surface and catalytic effects in the reactions of dissociated hydrogen-oxygen systems and the products condensed therefrom have been previously reviewed [65]. Semiokhin et al. [90] observed that the fraction of initial oxygen converted to H_2O_2 and the useful consumption of oxygen fell when a glass-metal (Al, Ni or Sn) ozonizer was used instead of an all-glass one. In these experiments at 1–3 atm the formation of water was found to be a catalytic process occurring on the metal electrode surface rather than one determined by the specific energy. Since aluminium gives an oxide which is very difficult to reduce, it is probable that the formation of water in the discharge is catalyzed by metal impurities in the aluminium, e.g. Fe, Zn or Cu.

4.1.1.3 Hydrazine Synthesis

The experiments [91,92] were made with ammonia using an all-glass ozonizer the surface of the inner electrode of which was deposited with a thin film of a metal. The metals used were Pt, Pd, Ni, Au and Ag. With recirculation of the gas mixture at a pressure close to atmospheric steady-state conversions into hydrazine as high as 13.5% were attained with Pd which is an increase in the N_2H_4 yield by a factor of ~4 (Fig. 18). The overall conversion of ammonia and the formation of N_2 and H_2 were almost independent of the metal coatings (Fig. 18) suggesting that hydrazine is not an intermediate in the decomposition of NH_3 into N_2 and H_2. Therefore, it seems that there are two independent reaction pathways: the formation of hydrazine which is a surface reaction that is dependent on the state of the electrode surface

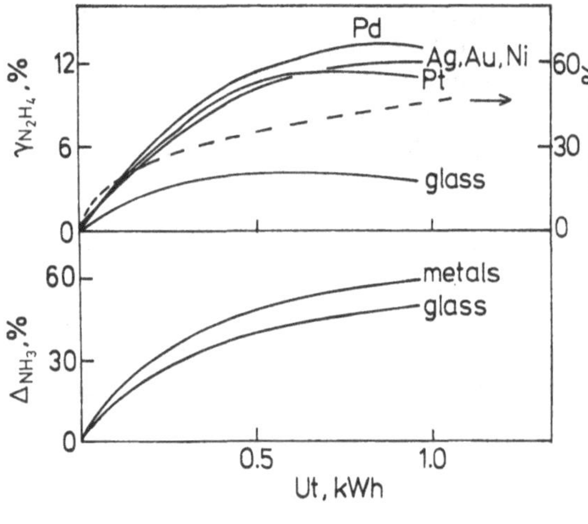

Fig. 18. Comparison of the effects of metals on the decomposition $\Delta\%$ of NH_3 (bottom) and on the formation $\gamma\%$ of N_2H_4 and N_2+H_2 (top) as final products in a barrier discharge at $p = 636$ torr and $i = 12.5$ mA. The catalysts used were Pt, Pd, Ni, Au and Ag

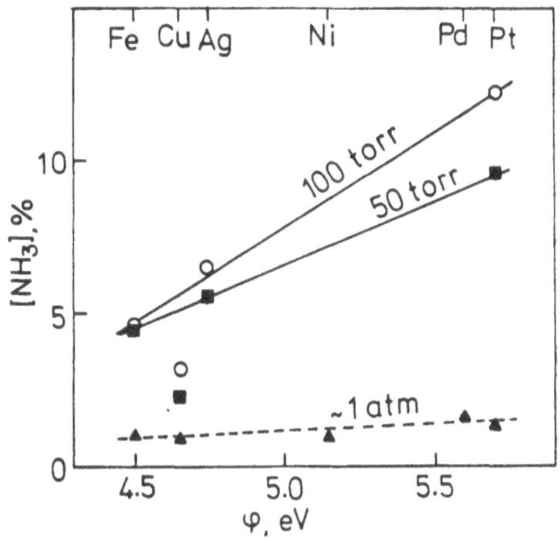

Fig. 19. Variation of the concentration of ammonia with the electron work function φ of the catalyst. Dashed line for experiments with metal wires wound over inner electrode of glass ozonizer [94]. Solid lines for experiments with metal gauze immersed in the plasma [95]

(actually a part of the walls of the discharge tube) and the decomposition of NH_3 into $N_2 + H_2$ which is a homogeneous reaction that is independent of the state of the electrode surface. Presumably the metal coatings retarded the decomposition of N_2H_4 in the gas phase by accelerating the recombination of hydrogen atoms.

The decomposition of ammonia in these experiments [93] was retarded by the addition of H_2 and accelerated by N_2 and Ar to about the same extent in the absence of catalysts, as in the presence of silver. This was taken as evidence that the decomposition is mainly a gas phase reaction. On the other hand, the formation of N_2H_4 was retarded by N_2 and H_2 to a considerably greater extent in the absence of catalysts than in the presence of Ag. Argon somewhat accelerated the formation of hydrazine in a reactor containing Ag, but retarded it in the absence of metals. Thus the formation of hydrazine appeared to be associated with the walls of the discharge tube. It was concluded that the inhibiting effect of H_2 and the accelerating actions of N_2 and Ar are spatial in character, probably consisting in the deactivation of excited NH_3 molecules by H_2 and in their activation by Ar and N_2.

4.1.1.4 Ammonia Synthesis

The effect of metal catalysts on the synthesis of ammonia at atmospheric pressure has been investigated [94] in a circulating system using an all-glass ozonizer supplied with 13 mA curremt at 500 Hz. The catalysts were wires made of Pd, Pt, Fe, Cu and Ni. About 21 m of 0.1 mm diameter wire was wound over the entire length of the inner electrode, the separation between the turns being $\simeq 1$ mm. Compared with clean glass an increase of ammonia concentration including the steady-state value $[NH_3]_{ss}$, was observed in the presence of the metals, the catalytic activity decreasing in the order Pd > Pt > Fe > Cu > Ni. In contrast to the conventional catalysis, here the activity is not estimated from the increase of the rate (rate constant) but from the increase of the steady-state concentration, which remains unchanged in the usual type of catalysis.

A linear relationship was found between the catalyst activity and the electron work function φ. A heterogeneous ionic mechanism (Sect. 4.1.1.5) was proposed [95,96] to account for the results: The N_2^+ ion which is discharged at the cathode displaces an electron and dissociates in the adsorbed state, forming N(ads) which reacts with H and H_2 from the gas phase:

$$N_2^+ \xrightarrow{\text{Surface}} 2N(ads) \tag{13}$$

$$N(ads) + H \rightarrow NH(ads) \xrightarrow{H_2} NH_3(ads) \rightarrow NH_3(g) \tag{14}$$

Possibly, for a larger φ the N_2^+ ion expends a greater part of its energy on the extraction of an electron from the lattice and displacement of another (which migrates to the plasma), the energy being otherwise consumed in the raising of temperature and desorption of atoms.

Table 2 lists the values of the kinetic constants calculated using the equation

$$[NH_3] = (k_1/k_2)\,[1 - \exp(-k_2Ut)_i] \tag{15}$$

Table 2. Steady-state concentrations of ammonia and the kinetic constants for its formation (k_1) and decomposition (k_2) at atmospheric pressure in the presence of catalysts on the electrode surface [94]

Catalyst	$[NH_3]_{ss}$,	Relative increase	10^3 dm^3 NH_3/Wh		k_1/k_2
	%	δ	k_1	k_2	
—	0.60	—	3.35	5.52	0.60
Pd	1.60	2.67	4.27	2.69	1.59
Pt	1.28	2.13	4.78	3.46	1.38
Fe	1.03	1.72	6.35	6.33	1.01
Cu	0.93	1.55	5.27	5.60	0.94
Ni	0.92	1.53	5.78	6.48	0.89

where the energy consumed $(Ut)_i$ was assumed to be the product of the discharge power U and the duration t of the experiment at constant current i. According to the kinetic analysis the catalysts investigated can be classified into two groups:
1) Pd and Pt and
2) Fe, Cu and Ni.

An increase of the rate constant for the formation of ammonia by a factor of ~1.5 with a simultaneous decrease of the decomposition rate constant by ~1.5–2, compared with the constants for clean glass, are characteristic of the first group. For the second group of metals, the formation rate constant increases by a factor of 1.5–2, while the decomposition rate constant does not change in the presence of Cu and actually increases slightly in the presence of Fe and Ni. It is interesting to note that Fe, which is the usual catalyst for the synthesis of NH_3, gives rise to the highest formation rate constant but not the highest yield.

In another series of experiments [94] palladized Pd and platinized Pt wires obtained by electrolyzing in $PdCl_2$ and H_2PtCl_6 solutions were found to be more active than

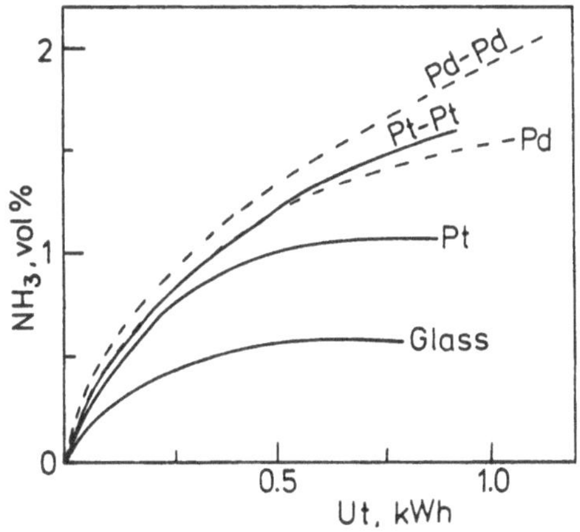

Fig. 20. Variation of the concentration of ammonia with the energy consumed for Pt, platinized Pt, Pd and palladized Pd catalysts [94]

Pd and Pt wires, respectively (Fig. 20), perhaps due to their large surface areas and to vigorous recombination of H atoms on these metals. On clean glass, the maximum concentration of NH_3 was reached for the stoichiometric $N_2 + 3\,H_2$ mixture; on palladized Pd for $N_2:H_2 = 2:3$ (Fig. 21). The displacement of the maximum concentration of NH_3 towards N_2 has been explained on the basis of an increase of the concentration of N_2^+ ions which are discharged on the surface of the metal and dissociate in the adsorbed state forming the N atoms necessary for NH_3 formation.

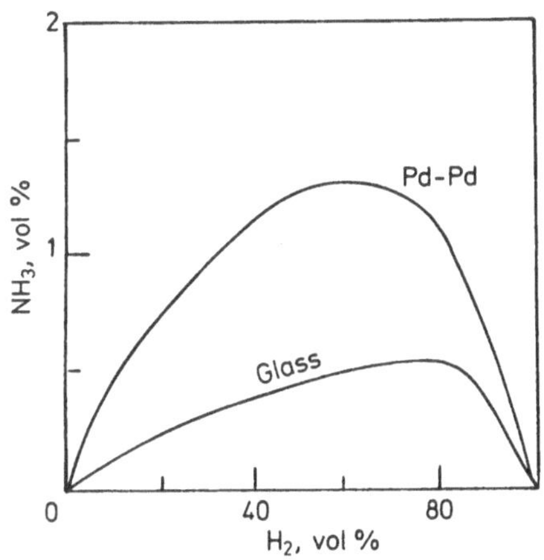

Fig. 21. Variation of $[NH_3]_{ss}$ with the composition of the initial nitrogen-hydrogen mixture for clean glass and palladized Pd as catalyst [94]

4.1.1.5 Heterogeneous Ionic Mechanism

The role of positive ions in plasma synthesis is best understood from studies of ammonia formation where a Pt catalyst in the form of a strip of gauze wound into a tight roll was fitted into the cavity of hollow water-cooled copper electrodes [97]. Figure 22 shows the variation of the concentration of ammonia with the specific

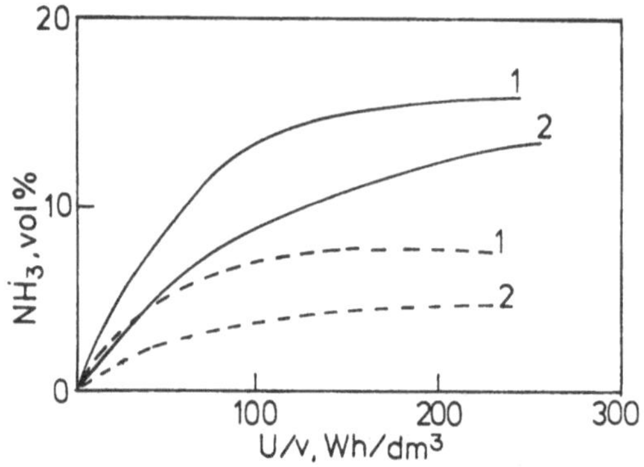

Fig. 22. Variation of the concentration of ammonia with specific energy of the glow discharge in the presence of Pt catalyst at the cathode (solid line) or anode (broken line) for electrode separations of (1) 10 and (2) 20 mm, respectively [97)]

energy U/v. When the catalyst was placed in the cathode the steady state concentration of NH_3 was the highest, which fact supports the heterogeneous ionic mechanism. When the catalyst was placed in the anode, the yield of ammonia decreased appreciably — by a factor of 2 and 3 at the inter-electrode separations of 10 and 20 mm, respectively. Perhaps with increase of the distance between the electrodes, the N atoms formed on the cathode find it more difficult to reach the catalyst-anode.

In the oxidation of nitrogen [98,99)] Pt and Fe catalysts placed on the anode has no significant effect on $[NO]_{ss}$. If the catalysts were placed in the cathode the yield of NO increased appreciably (Table 3). Similarly, studies [100,198)] of the synthesis of hydrocarbons from graphite and hydrogen in a dc glow discharge established that bombardment of the graphite surface by energetic ions enhances the reaction rate. Similar applies also to other systems (see Sect. 5.2). In the gasification of graphite and coal by CO_2 [101)] the highest steady state concentration of CO was obtained in experiments with the carbon placed on the cathode rather than on the anode.

Table 3. Comparison of $[NO]_{ss}$ for catalyst on the same electrode used as cathode or anode [98,99)]

p	No catalyst	Cathode		Anode	
torr		Pt	Fe	Pt	Fe
50	0.19	4.50	2.67	0.58	0.51
75	1.53	4.42	2.77	1.75	1.81
100	2.50	5.49	4.07	3.19	2.93
	4.21[a]	—	6.49[a]	—	5.12[a]

[a] Values for discharge current of 200 mA; otherwise 100 mA was used.

4.1.2 Effects Due to Materials in the Inter-electrode Space

4.1.2.1 Ozone Synthesis

The rate of formation of ozone in an oxygen plasma at atmospheric pressure increased when the ozonizer was packed with materials such as soda glass, TiO_2, 2 MgO/TiO_2, $BaTiO_3$ and 95 $BaTiO_3$/5 SnO_2 [102)]. A linear relationship was found

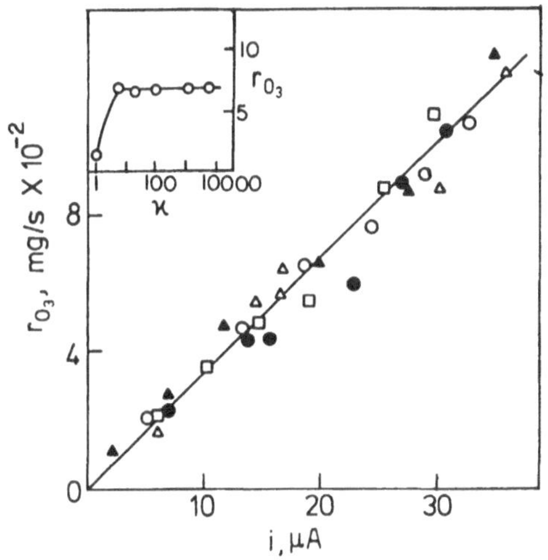

Fig. 23. Rate of ozone formation (r_{O_3}) under various packings against pulse current i. □, ○, ●, △, and ▲ for soda glass, TiO_2, 2 MgO/TiO_2, $BaTiO_3$ and 95 $BaTiO_3$/5 SnO_2, respectively. Inset shows r_{O_3} versus dielectric constant ϰ at i = 21 μA [102]

between the discharge current and the ozone formation rate with and without packing (Fig. 23) and it is independent of the value of the dielectric constant of the packing materials (Fig. 23). The ozone concentration was a maximum at a certain value of the dielectric constant (<10 for soda glass) and was independent of the dielectric constant (>10 for all other materials used) at the higher values of the current (inset Fig. 23).

4.1.2.2 Hydrogen Peroxide Synthesis

Morinaga [103] found that packings such as soda glass, MgO, TiO_2 and $BaTiO_3$ destroyed H_2O_2 whereas B_2O_3 and H_3BO_3 did not affect the peroxide yields from dissociated oxygen-hydrogen systems at atmospheric pressure in borosilicate (Pyrex) ozonizer tubes. This suggested that catalytic effects and not surface-to-volume ratio is significant in the decomposition of H_2O_2. In low pressure systems Brunet et al. [104] have reported similar surface and packing effects on the formation of H_2O_2 from dissociated water vapor. At first it was suggested [105,106] that the oxygen evolution from condensed products in these systems was due entirely to peroxide decomposition by impurities such as silica and metal particles from the discharge carried over by the gas stream. Subsequently, it was proven [107,108] that neither catalytic nor stoichiometric decomposition of peroxide could account for the evolved oxygen.

4.1.2.3 Hydrazine Synthesis

When solid NaOH or KOH was introduced [109] into the space between the electrodes in $\simeq 5$ mm lumps and a discharge in ammonia was operated at 5 torr and 5 mA the hydrazine content increased by a factor of about seven (Fig. 24). However, the continuous operation of the discharge caused the catalyst to melt. With intermittent operation of the discharge [110] the catalyst did not melt and the degree of conversion into hydrazine reached $\simeq 15\%$, double that obtained with a continuous discharge (Fig. 24). The effect was explained by assuming that the formation of hydrazine

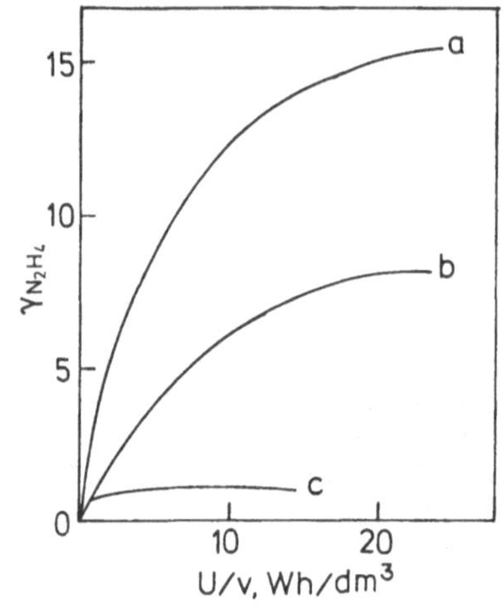

Fig. 24. Dependence of the degree of conversion γ of ammonia into hydrazine on specific energy (i = 5 mA, p_{NH_3} = 5 torr) for (a) intermittent and (b) continuous discharges with NaOH as catalyst. Curve (c) is for continuous discharge in the absence of a catalyst [110]

involves the reaction between ammonia molecules sorbed on the catalyst surface and the NH radicals in the gas phase, which were observed spectroscopically:

$$NH_3\ (ads) + NH \rightarrow N_2H_4\ (ads) \quad (16)$$

Surely the mean surface coverage of the catalyst with NH_3 molecules is higher during an experiment with an intermittent discharge than with a continuous discharge under similar conditions.

Spectroscopic investigations [43,44] also showed that the introduction of the solid alkali did not lead to an anomalous population of rotational levels in NH but raised the vibrational temperature from $\simeq 3000$ K to $\simeq 4000$ K. It was suggested that the recombination of ions, presumably N_2^+, on the catalyst surface leads to an increase in the number of nitrogen molecules at high levels of vibrational excitation. However, the location of NaOH catalyst at the cathode or anode does not influence significantly either the decomposition of ammonia or the synthesis of hydrazine (Fig. 25) [111]. This suggested that ions do not play a significant role in the synthesis of hydrazine on the surface of solid alkali.

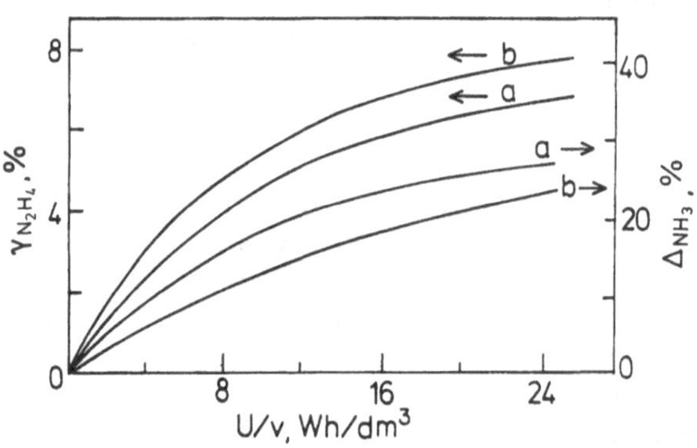

Fig. 25. Variation of the degree of conversion Δ of ammonia and conversion γ into hydrazine with specific energy (i = 10 mA, p = 3 torr) for the catalyst (a) on the anode and (b) on the cathode [111]

The replacement of the basic alkalis by acidic silica gel reduced the yield of hydrazine almost to zero [109]. Therefore, the basic properties of the catalyst are the main factors in its action to increase the hydrazine yield. Thus the replacement of NaOH by KOH does not alter the catalyzed reaction significantly.

In a low pressure (1–40 mbar) plasma produced by an inductively coupled rf discharge Rapakoulias et al. [3] found that the rate of decomposition of ammonia to N_2 and H_2 was increased very much by the presence of W or Mo. Further, the catalytic effect increased with increasing degree α of nonequilibrium in the plasma (Fig. 26):

$$\alpha = (T_v - T_t)/(T_v + T_t) \tag{17}$$

where T_v and T_t are the vibrational and translational temperatures in the plasma. A spectroscopic analysis of the gas in the close vicinity of the catalyst surface showed the existence of a boundary layer in which the vibrational excitation of N_2 is significantly higher than in the bulk plasma. It was concluded that in this boundary a v-v transfer is occurring from the desorbing N_2 to the adsorbing NH_3. This feedback transfer surely will increase the efficiency of dissociative adsorption of NH_3 leading to the observed catalytic effect.

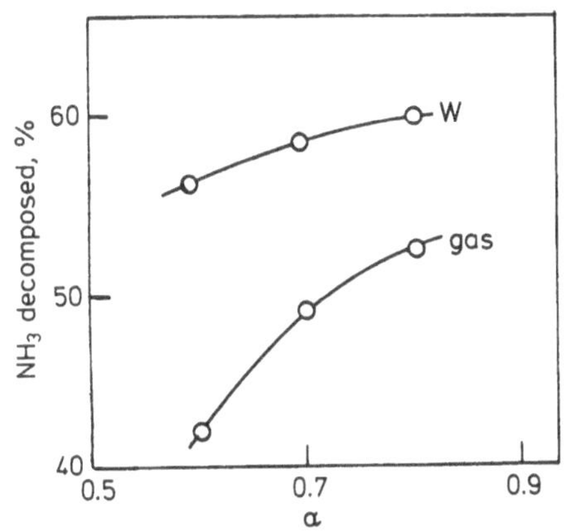

Fig. 26. Variation of the decomposition rate of NH_3 as a function of the degree α of nonequilibrium in the plasma [3]

4.1.2.4 Ammonia Synthesis

The catalytic effects of Pt, Ag, Fe and Cu gauze formed into rolls and introduced directly into the interelectrode space, where they were exposed to the discharge plasma in $N_2 + H_2$ mixtures, have been investigated [27,95,96].

Some of the experiments [95,96] were done at pressures of 50 and 100 torr and gas flow rates 1–50 $dm^3\ h^{-1}$. Under the influence of the discharge alone, a steady-state concentration of ammonia was attained fairly rapidly at relatively low values of U/v (Fig. 27), but in the presence of a catalyst a steady state was not attained presumably due to the experimental conditions which did not provide sufficiently high specific energies. Therefore, for experiments with catalysts the steady-state concentrations of NH_3 shown in Fig. 27 were obtained by circulating the gas in a closed system.

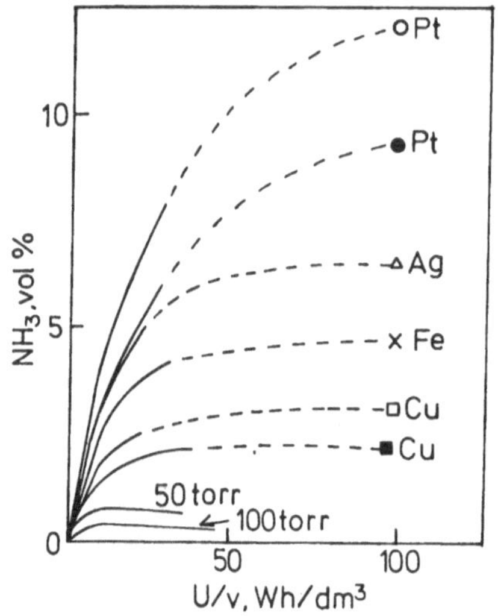

Fig. 27. Variation of the ammonia concentration with specific energy at a pressure of 50 torr (curves leading to open points) and 100 torr (curves leading to solid points) for different catalysts. Dotted parts were not determined experimentally. Points from circulation experiments; solid lines from flow experiments [96]

Table 4. Steady-state concentrations of ammonia and the kinetic constants for its formation (k_1) and decomposition (k_2) at low pressures in the presence of catalysts in the interelectrode space [95] [a]

Catalyst	$[NH_3]_{ss}$, %	Relative increase δ	dm³ NH_3/Wh k_1	k_2	k_1/k_2
—	0.75	—	0.106	0.141	0.75
	0.36	—	0.139	0.356	0.39
Pt	12.22	16.3	0.413	0.034	12.15
	9.6	26.6	0.311	0.032	9.72
Ag	6.5	8.68	0.371	0.057	6.51
	5.55	15.4	0.198	0.036	5.5
Fe	4.6	6.15	0.354	0.077	4.6
	4.53	12.6	0.158	0.035	4.52
Cu	3.15	4.2	0.235	0.073	3.22
	2.3	6.4	0.203	0.089	2.29

[a] Two sets of values given correspond to initial pressures of 50 and 100 torr, respectively.

Table 4 shows that the activity of the catalysts varied in the sequence Pt > Ag > Fe > Cu. The rate constante for the formation (k_1) and the decomposition (k_2) of ammonia were calculated using eq. 15. In the presence of catalysts k_1 exceeded k_2, sometimes by an order of magnitude, but without catalysts k_1 was significantly lower than k_2. This indicated that the displacement of the steady state was due to different effects of the catalyst on the rates of formation and decomposition of ammonia in a plasma. The linear relation (Fig. 19) between catalyst activity and the electron work function φ was adduced as evidence in support of a heterogeneous ionic mechanism (Sect. 4.1.1.4).

In a recent investigation [27] at a lower pressure ($\simeq$ 10 torr) using an Fe catalyst a relative increase (4–5) in NH_3 yield comparable to those reported in Table 4 for the higher pressures was found. Further, it was found that the temperature of the

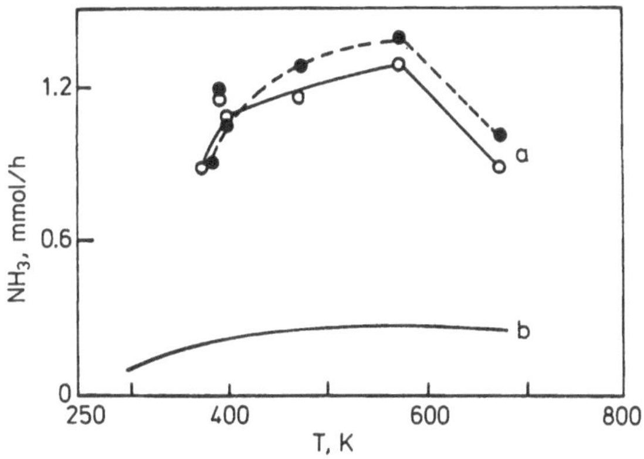

Fig. 28. Effect of Fe catalyst temperature on NH_3 yield in a discharge at 8 mA. (a) Fe in the plasma zone; (b) plasma gas passed over Fe. Open points for $p_{N_2} = 3$ torr and $p_{H_2} = 6$ torr; closed points for $p_{N_2} = 6$ torr and $p_{H_2} = 3$ torr [27]

catalyst is an important parameter in optimizing the NH_3 yield (Fig. 28). Analysis of the plasma emission before and after the catalyst zone revealed an increase of NH and NH_2 concentrations suggesting a heterogeneous radical mechanism leading to ammonia formation:

$$NH + H_2 \rightarrow NH_3^* \tag{6}$$

The low steady-state concentrations of ammonia in the discharge in experiments without a catalyst was taken as evidence that the large excess energy of NH_3^* molecules is removed more effectively on metallic surfaces than homogeneously by triple collisions.

Introduction of solid KOH in the space between the electrodes increased the concentration of NH_3 by a factor of 18 at $U/v = 240$ Wh dm^{-3} [112]. In these experiments at $\simeq 3$ torr and 20–40 mA the melting of the catalyst was avoided by using an intermittent discharge, which operated for 1 s with a 4 s interruption. Certainly, a part of the increase in NH_3 yield should be attributed to pulsing of the discharge which is known to minimize product decomposition. Further, decomposition reactions such as

$$NH_3 + H \rightarrow NH_2 + H_2 \tag{7}$$

were probably suppressed because H atoms on the surface of the hydrated alkali may be eliminated by formation of hydrated electrons:

$$H + OH^- (aq) \rightarrow e (aq) + H_2O \tag{18}$$

4.1.2.5 Nitric Oxide Synthesis

The catalytic oxidation of nitrogen

$$N_2 + O_2 \rightleftharpoons 2\,NO \tag{1}$$

has been studied in a glow discharge with electrodes [113–115] and electrodeless discharges [116, 117].

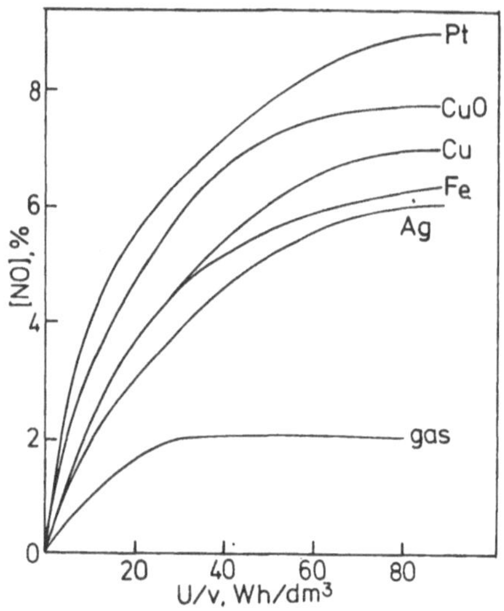

Fig. 29. Dependence of [NO] % on the specific energy at a pressure of 50 torr for a stoichiometric mixture of N_2 and O_2 [114]

The glow discharge experiments [113–115] were performed in a flow-circulation system at a current of 35 mA. The catalyst consisted of rolled up metallic gauze of Pt, Cu, Ag or Fe and CuO placed in the positive column of the plasma. For a stoichiometric mixture of N_2 and O_2 the variation of the percentage of NO with specific energy is shown in Fig. 29. In terms of their activities the catalysts exhibited the following sequence: Pt > CuO > Cu > Fe > Ag. The effectiveness of the metal gauze appeared to relate linearly to the electron work function φ (Fig. 30) and to support a heterogeneous ionic mechanism involving N_2^+. The rate constants for the forward (k_1) and reverse (k_2) reactions calculated using eq. (2) are given in Table 5. It is readily seen that the formation rate constant k_1 increases in the presence of the catalysts, while the decomposition constant k_2 diminishes, which leads to an increase of the ratio $k_1/k_2 = [NO]_{ss}$.

The relative increases in the steady state concentration of NO for different reaction mixtures given in Table 6 show that a stoichiometric mixture over Pt

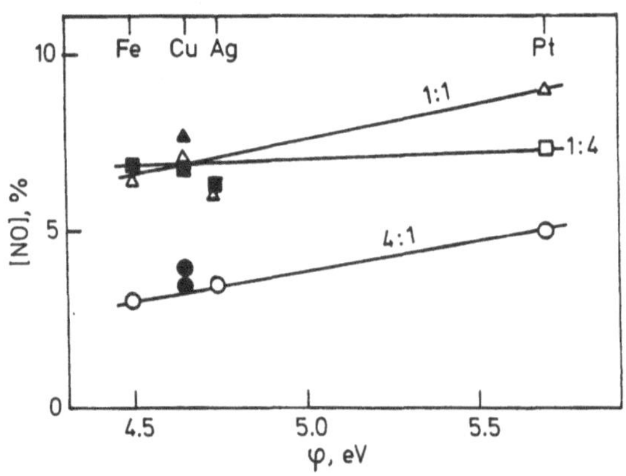

Fig. 30. Variation of $[NO]_{ss}$ with electronic work function φ of the catalyst for different compositions of N_2 and O_2. Shaded points represent values of $[NO]_{ss}$ for oxide catalysts at φ for the corresponding metal. □ for $N_2:O_2 \simeq 1:4$; ○, 4:1 and △, 1:1 mixtures [113–115]

Table 5. Steady-state concentrations of nitric oxide and kinetic constants for a stoichiometric nitrogen-oxygen mixture at 50 torr in the presence of catalysts in the interelectrode space [114]

Catalyst	$[NO]_{ss}$, %	dm³ NO/Wh		k_1/k_2
		k_1	k_2	
—	2.10	0.17	0.08	2.1
Pt	8.98	0.44	0.049	8.98
CuO	7.80	0.38	0.049	7.78
Fe	6.40	0.25	0.039	6.4
Ag	6.10	0.22	0.036	6.11
Cu	7.03	0.27	0.038	7.03

is the best choice for producing maximum NO under plasma catalytic conditions. Although the noncatalyzed reaction was found to be pressure dependent, the catalyzed reaction was independent of pressure. This was explained by the fact that the reaction in the discharge in the presence of catalysts is determined mainly by the number of N_2^+ ions discharged at the cathode, which is independent of the pressure. Since Townsend's second coefficient is small ($\simeq 0.02$) under the conditions of these experiments the number of N_2^+ ions undergoing discharge can be found directly from the current. Table 7 shows that the number of NO molecules formed for each N_2^+ ion in the presence of the catalysts is higher than in the experiments without catalysts and increased in the same sequence as the activities of the catalysts.

To explain the dependence of NO yield on composition a mechanism was proposed [115] in which N_2^+ ions play an important role. In the case of air the mechanism included reactions for the *decomposition* of NO: [118]

$$NO + e^- \rightarrow NO^- \tag{19}$$

$$NO^- + NO \rightarrow (NO)_2^- \tag{20}$$

$$(NO)_2^- \rightarrow N_2 + O_2 + e^- \tag{21}$$

With stoichiometric mixture there is an increase of the oxygen concentration in the mixture which prevents the decomposition of NO. This is evident from the analysis of the kinetic constants (Table 6): k_1 does not change on passing from air to a stoichiometric mixture, while k_2 greatly diminishes, that is, the reaction

$$(NO)_2^- + O_2 \rightarrow 2\,NO + O_2^- \tag{22}$$

is preponderant over the reaction:

$$O_2 + e^- + M \rightarrow O_2^- + M \qquad (M = O_2 \text{ or } N_2) \tag{23}$$

In the case of the 1:4 nitrogen-oxygen mixture reaction (23) was preferred over reaction (22), since a large amount of O_2 in the mixture promotes a more complete utilization of the N_2^+ ions, which can be seen from the increase of k_1 for the $N_2:O_2 \simeq \simeq 1:4$ mixture compared with the stoichiometric mixture and air.

Table 6. Activities of catalysts[a] placed in the interelectrode space and kinetic constants[b] for different compositions of nitrogen-oxygen mixtures[113–115)]

Catalyst	p	N_2/O_2 = 4			1			0.25		
	torr	δ	k_1	k_2	δ	k_1	k_2	δ	k_1	k_2
Pt	25	—	—	—	7.36	0.40	0.045	6.3	0.72	0.1
	50	3.97	0.48	0.096	7.28	0.44	0.049	4.0	0.72	0.1
	100	2.75	0.50	0.10	—	—	—	—	—	—
CuO	50	3.17	0.32	0.08	3.71	0.38	0.049	3.73	0.57	0.085
	100	1.92	0.35	0.1	3.67	0.36	0.046	2.25	0.57	0.085
Ag	50	2.82	0.32	0.09	2.90	0.22	0.036	—	—	—
	100	1.91	0.30	0.086	—	—	—	—	—	—
Fe	50	2.46	0.31	0.1	3.05	0.25	0.039	—	—	—
	100	1.62	0.29	0.1	—	—	—	—	—	—
FeO	25	—	—	—	—	—	—	6.0	0.56	0.082
	50	—	—	—	—	—	—	3.78	0.56	0.082
Ag_2O	50	—	—	—	—	—	—	3.48	0.47	0.074
Cu	50				3.35	0.27	0.038			

[a] Given as δ the relative increase in $[NO]_{ss}$;
[b] Given as dm^3 NO/Wh.

Table 7. Number of NO molecules formed for each N_2^+ ion discharged at the cathode with the catalyst in the interelectrode space [113–115]

Catalyst	p =	25 torr			50 torr			100 torr		
	N_2/O_2 =	4	1	0.25	4	1	0.25	4	1	0.25
—		1.27	2.07	1.13	2.9	2.6	2.12	4.3	2.1	2.67
Pt		—	5.15	9.1	6.85	6.36	9.7	7.64	—	—
CuO		—	—	—	4.6	5.5	8.0	5.9	5.4	8.7
Fe		—	—	7.26	4.2	3.67	8.18	4.7	—	—
Ag		—	—	—	4.13	3.18	6.53	4.6	—	—
Cu		—	—	—	—	3.9	—	—	—	—

A steady-state treatment of the mechanism [115] showed that in the cases of air and stoichiometric mixtures the rate of formation of NO is proportional to the number of N_2^+ ions undergoing discharge on the cathode, that is, to the current. This is supported by the calculation of the number of NO molecules, formed for each N_2^+ ion discharged at the cathode (Table 7). The higher yields of NO molecules with 1:4 mixture suggested that in this case the rate also depends on O_2 concentration.

In inductively coupled rf discharges [116,117] at 0.5–1.5 kW WO_3 and MoO_3 exhibited strong catalytic effects in the oxidation of N_2: Nitrogen fixation attained 19% compared with 8% without the catalyst. Analysis of experimental results showed chemisorption of vibrationally excited N_2^* followed by surface reaction with mobile O of the catalyst to be the rate determining step of the catalytic process. The non-equilibrium of the plasma accelerated this catalytic mechanism in comparison with thermal systems. With these catalysts also a significant change in NO yield occurred with mixtures containing more than stoichiometric amounts of oxygen (Fig. 31).

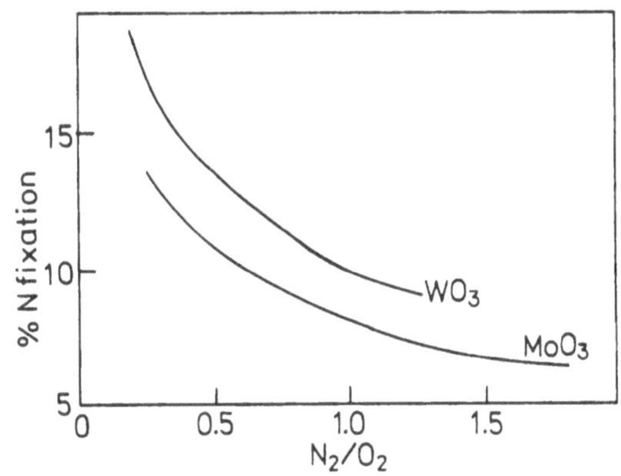

Fig. 31. Variation of % fixation of N_2 as NO with gas composition for WO_3 and MoO_3 catalysts [116]

4.1.2.6 Hydrogen Cyanide Synthesis

The direct fixation of nitrogen by its reaction with CH_4 in a plasma has been studied extensively [119]. At low pressure hydrogen cyanide, acetylene and hydrogen are produced [120]. A kinetic study [121] of the reaction in the pressure range 5–30 torr showed that the slow step is represented by the rate of dissociation of N_2 in the discharge. It was proposed [122] that thermal dissociation occurs from excited vibrational-rotational levels, the population of which is maintained by the electric field at values much higher than those corresponding to thermal equilibrium.

In inductively coupled rf discharges at 1–40 mbar N_2 fixation as HCN increased very much in the presence of W, Ta and Mo [3,123a]. Molybdenum was most efficient and enhanced HCN production by a factor of 50% or more. The catalytic effect increased with increasing nonequilibrium degree α which is defined by eq. 17 (Fig. 32). A catalysis mechanism was proposed in which the lowest molecular metastable state of $N_2(A^3\Sigma_u)$ or the N_2 fundamental (vibrationally excited by the discharge) gain, by chemisorption on the metal, the excess energy necessary for their dissociation [123b].

Hydrogen cyanide is also produced in the plasma reactions of ammonia with methane [124] and with benzene [64]. Bundles of Ni inserted in a microwave discharge [64] in mixtures of benzene and ammonia at a pressure of 40 mbar lowered the total conversion of benzene, but it had little effect on the conversion of ammonia. The nickel also altered the distribution of reactant products: the formation of aniline was enhanced but the formation of HCN and C_2H_2 were reduced.

The presence of hydrocarbons invariably leads to a brownish-yellow polymer deposit on the plasma walls. Laser action ($\lambda = 337$ μm) is observed in tubes containing this type of deposit if the discharge is maintained in pure H_2, H_2O or NH_3. Schoetzau and Vepřek [125] have explained the experimentally observed gain of HCN lasers in terms of the combined effects of volume processes and wall reactions — the latter include the formation of excited HCN by the reaction of H atoms with the deposit formed by discharge-activated polymerization of hydrocarbons in the laser tube.

4.1.2.7 Reduction of Carbon Dioxide By Hydrogen

The hydrogenation of CO_2 is the basis of the catalytic purification of technological gases. Therefore, during recent years increasing attention has been paid to the plasma reactions of mixtures of CO_2 and H_2. Using a manometric technique and various proportions of CO_2 and H_2 at pressures of 50–600 torr and discharge currents

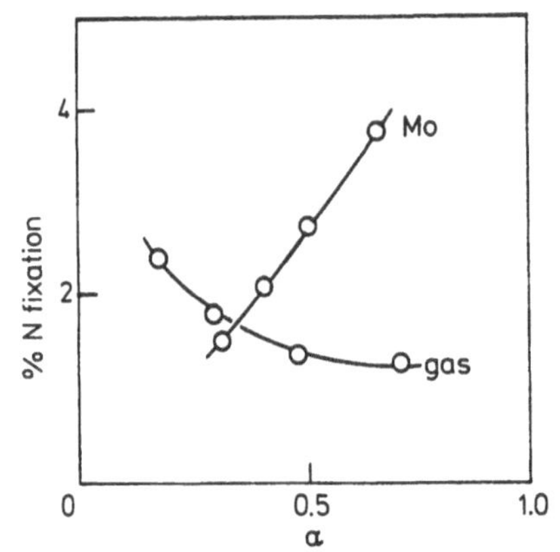

Fig. 32. Variation of rate of HCN synthesis as function of the degree α of nonequilibrium in the plasma [3]

of 100–300 mA in a circulatory water-cooled system fitted with brass electrodes Eremin et al.[126] found that the reaction occurs in two stages

$$CO_2 + H_2 = CO + H_2O \quad (I)$$

$$CO + H_2 = C + H_2O \quad (II)$$

both of which were describable by second order equations. The ratios of the rate constants k_I/k_{II} suggested that the second stage occurs more slowly than the first, especially at high pressures. The difference in the rates of the two stages is apparently due to the lower bond strength of CO_2 (526 kJ mol^{-1}) compared with CO (1070 kJ mol^{-1}). It must, however, be noted that carbon introduced into a water vapor discharge under comparable conditions is eroded and CO and even CO_2 are produced in such discharges[65].

By narrowing the diameter of the discharge tube from 20 to 3 mm a four-fold acceleration of the reduction of CO_2 was accomplished[127]. This was interpreted on the basis of the importance of surface reactions in the process. However, Vepřek (unpublished) has observed that the steady state of the reaction

$$CO_2 = CO + O$$

shifts to the right-hand-side with increasing discharge current density at constant tube diameter and that at a sufficiently high current density even CO decomposes in the plasma[163]. The presence of a tight roll of iron gauze in the plasma zone gave an acceleration factor of 32[128]. When grains of mixed catalysts (generally used for NH_3 synthesis) of the following compositions:

	% Fe_2O_3	FeO	Al_2O_3	CaO	K_2O	SiO_2	MgO
I	61	30	4	3	1	1	0
II	51.81	31.84	7.47	2.32	1.18	0.84	4.54

and activated in hydrogen were placed in the plasma region on an iron support grid the acceleration factors were 143 and 125, respectively[138]. Such substantial accelerations make the rate of the first step comparable to that of the second step. In successive experiments the acceleration factor remained almost constant so that the deposition of reaction products (C and H_2O) on the catalyst apparently did not deactivate the catalyst.

4.1.3 Effects Due to Materials on Plasma Walls

4.1.3.1 Hydrazine Synthesis

Devins and Burton[29] reported an increased yield of hydrazine by about a factor of three if platinum-coated walls were used in dc discharges in ammonia. Rubtsova and Eremin[129] also found significant catalytic effects with Pt and thin coatings of a number of other metals deposited on the walls of a quartz tube 22 mm in diameter (Fig. 33). When the current was raised from 1 mA to 11 mA, the relative activity of Pd increased but that of Ni was unaffected. With increase in temperature the

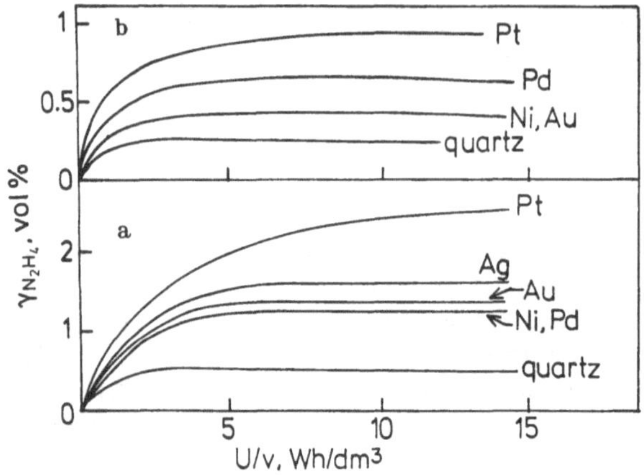

Fig. 33. Variation of the degree of conversion γ into hydrazine with specific energy at a current of (a) 1 mA and (b) 11 mA [129]

transition metals give up the hydrogen which entered into their crystal lattices. Palladium is almost free of hydrogen at ≃300 °C, but nickel only at about 700 °C. Clearly an increase of wall temperature due to the increase of the current in the range studied is insufficient to remove hydrogen from the nickel and therefore its activity remained low as before.

Although the metals greatly increased the extent of formation of hydrazine they caused little effect on the decomposition of ammonia [129]. However, the addition of H_2 and N_2 in the presence of Ag inhibited both the decomposition of ammonia and the synthesis of hydrazine, N_2 showing a more pronounced inhibiting effect [129]. Since Ar had no influence under the same conditions, but, like N_2 and H_2, accelerated the decomposition of NH_3 and retarded the formation of hydrazine in the absence of metals [130], suggesting that the added gases decomposed both NH_3 and N_2H_4 in the bulk phase, behaving as energy catalysts of the decomposition process. In the presence of Ag, surface processes probably played the main role which accounts for the inertness of Ar.

4.1.3.2 Ammonia Synthesis

Copper, Ni, Fe and Pt deposited on the discharge tube walls by cathodic sputtering at a low pressure, increased the stationary concentrations of ammonia by factors of 1.5–2.0 compared with an uncoated silica surface (Table 8) [131]. The experiments

Table 8. Steady-state concentration of ammonia as a function of discharge current for various metal coatings in the discharge tube at 50 torr [131]

Catalyst		$[NH_3]_{ss}$, %			Relative increase		
	i (mA) =	35	100	225	35	100	225
—		0.94	0.81	1.0	—	—	—
Pt		1.35	1.63	1.98	1.44	2.02	1.8
Cu		1.24	1.32	—	1.32	1.63	—
Fe		—	1.34	1.71	—	1.66	1.71
Ni		1.03	1.30	1.5	1.1	1.61	1.5

were performed with stoichiometric mixtures of N_2 and H_2 at $\simeq 50$ torr in a discharge tube of 3 mm diameter. The data given for coatings in Table 8 are to be compared with those given in Table 4 for catalysts placed in the interelectrode space. The latter location for the catalyst appears to have a distinct advantage with respect to product yield. At lower pressures ($\simeq 1$ torr) and currents ($\simeq 15$ mA) increasing the temperature of the Pt-coated wall in a wider tube (d = 12 mm) increased the rate of formation of NH_3 to a maximum at $\simeq 600$ K (Fig. 34) [132]. This is to be compared with a similar temperature effect observed for a Fe catalyst placed in the plasma zone (Fig. 28) [27].

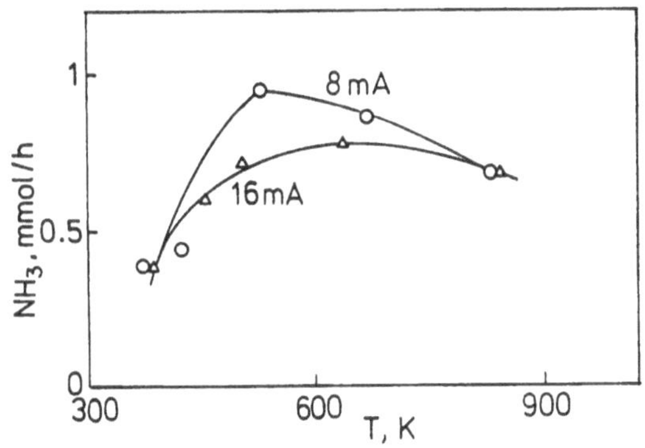

Fig. 34. Variation of NH_3 yield as a function of temperature of the Pt-coated discharge tube walls [132]

4.2 Catalytic Effects Occurring in the Spatial Afterglow

In some discharge-flow systems notably O_2 [133], H_2O [65, 134, 135], H_2O_2 [136, 137], and $N_2 + H_2$ [138] the yields of products were found to depend on reaction time of the dissociated gases in the spatial afterglow rather than in the plasma itself. When the surface-to-volume (S/V) ratio of the spatial afterglow region was increased in experiments using dissociated water vapor [139], the yield of hydrogen peroxide in a cold trap at 77 K increased to a maximum value and then decreased (Fig. 35). Only in one

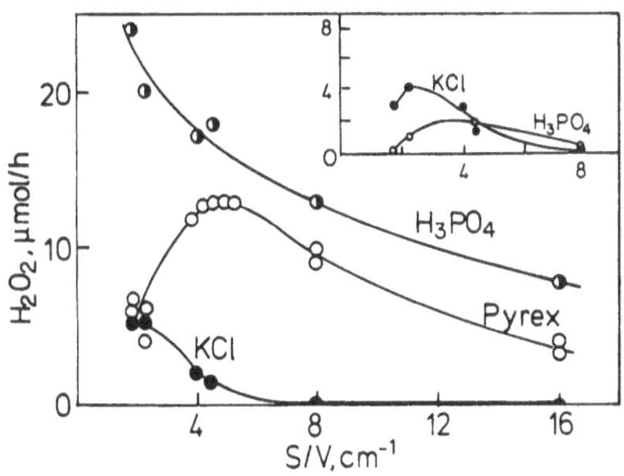

Fig. 35. Yield of H_2O_2 as a function of surface-to-volume (S/V) ratio of uncoated (Pyrex), H_3PO_4-coated, and KCl-coated after-glow regions. Inset shows the H_2O_2 yield as a function of S/V ratio of uncoated (Pyrex) afterglow region for coated cold traps at 77 K [139]

investigation [65] the plasma was produced using discharges with internal electrodes so that these dependencies of the product yield were ascribed to non-catalyzed processes.

4.2.1 Effects Due to Wall Coatings

In experiments using dissociated water vapor [139] coating the glass walls of the spatial afterglow region with H_3PO_4 increased the H_2O yield considerably at all S/V ratios studied, but coating with KCl decreased the H_2O_2 yield very much and practically no H_2O_2 was obtained at $S/V > 8\ cm^{-1}$ (Fig. 35). When the same region was jacketted with either ice-water (0 °C) or Dry Ice-acetone (—80 °C) and water was deposited on the walls prior to the discharge, the H_2O_2 yield decreased by about 20–25%. These effects have been explained in terms of the degree of "wetness" of the surfaces and correlated with the acidity or alkalinity of surfaces. For example, on acidic surfaces a proton is donated and on alkaline surfaces an electron is donated to the reactive species such as O_2H:

$$O_2H + H^+\text{-surface} \rightarrow {}^1/_2\, H_2O_2 + {}^1/_2\, O_2 + H^+\text{-surface} \tag{24}$$

$$O_2H + e^-\text{-surface} \rightarrow {}^1/_2\, H_2O + {}^3/_4\, O_2 + e^-\text{-surface} \tag{25}$$

In the absence of work using labeled coatings there has been no verification of the mechanism involved in these surface exchange processes.

There are few investigations of heterogeneous catalysis in the afterglow region. The effect of platinum sputtered on to the glass walls of the afterglow region of a microwave discharge in $N_2 + H_2$ mixtures has been reported [140]. Figure 36 shows the mol% conversion of N_2 to NH_3 as a function of the reaction time t of N_2 in the spatial afterglow. It also shows the catalytic effects on the $N + H_2$ reaction. At $t > 1$ s, the catalytic effects were temperature dependent, as it was indeed the case [27] for Pt placed in the glow itself. Increasing the pressure and/or concentration of N_2 in the gas mixture increased the yield of NH_3. For example, a mixture containing 12 mole % N_2 at 0.1 torr yielded up to 13 mole % NH_3. As the pressure was increased

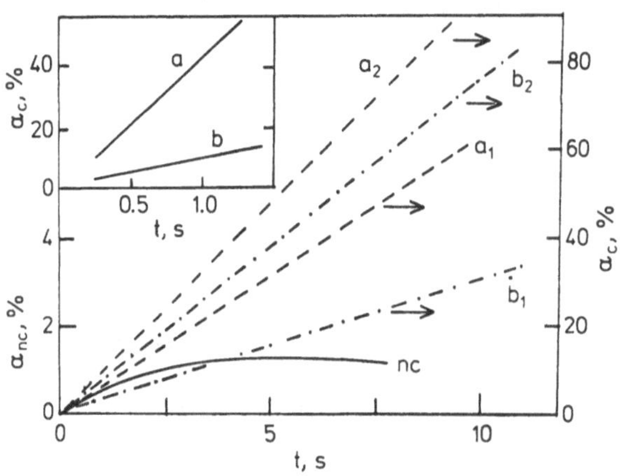

Fig. 36. Catalyzed (c) and non-catalyzed (nc) N_2 conversion α (mol %) to NH_3 versus reaction time (t) in the spatial afterglow of (a) $N_2 + H_2$ and (b) $N + H_2$ reaction at (1) 300 K and (2) 633 K [140]

to 0.7 torr the NH_3 yield increased to $\simeq$27 mole %. This is very much greater than the best value (12%) reported by Eremin et al. [95,96] for heterogeneous catalysis by Pt in a glow discharge.

4.2.2 Effects Due to Catalysts Placed in the Afterglow

When an Fe catalyst was placed in the afterglow of a $N_2 + H_2$ plasma more ammonia was obtained, but not as much as that was obtained with the same catalyst placed in the plasma zone (Fig. 28) [27]. Significant nitriding of iron is known to occur in low pressure discharges [141], but it was not known whether nitriding was responsible for the high yields. However, catalysts such as CuO and Fe_2O_3 produced more nitric oxide when placed in the afterglow than in the plasma zone [142]. Further, the NO yields were greater with the oxide catalysts than the respective metals [142]. In this connection it is interesting to note that treatment of Cr_2O_3 in hydrogen (reductive) and argon (inert) plasmas showed strong warm-up with simultaneous formation of nitrogen oxides when exposed to air [143,144].

In connection with the synthesis of methane and other organic substances plasma reduced mixtures of CO_2 and H_2, that is, CO and H_2O have been passed over a number of mixed catalysts [145] previously reduced in a stream of H_2 for several hours at 420–440 °C. Figure 37 shows a typical kinetic curve from such an experiment. The upper section *ab* of the curve corresponds to the plasma reaction, the horizontal section *bc* corresponds to the conditioning at 150 °C of the catalyst, and the vertical section *de* corresponds to the afterglow stage of the reaction which is rapid. Note that an appreciable acceleration begins immediately after the temperatures was raised to 185–190 °C (Section *cd* of the curve). Since the process is highly exothermic, much heat is evolved, the temperature rises rapidly and the catalyst is heated. To stop the reaction the catalytic reactor was cooled to 185 °C (Section *ef* of the curve).

A number of these experiments [145] at 800 torr and 75 mA using a Co—Zr catalyst showed that in the temperature range above 190 °C the catalytic reaction produced chiefly methane (Table 9) and to some extent liquid products — ethanol, hydrocarbons, formaldehyde and water — collected in a trap at 195 K. A partial

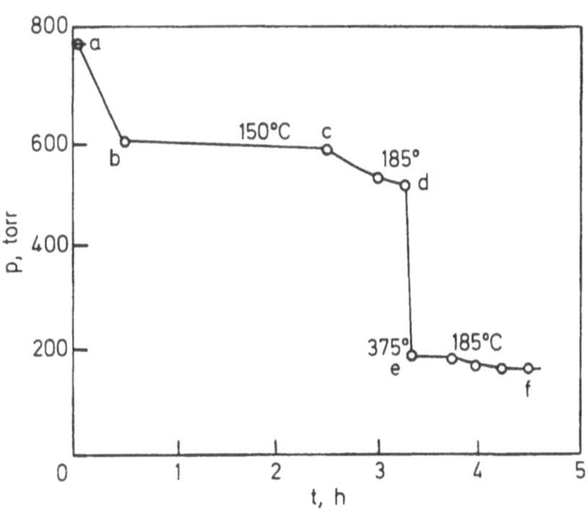

Fig. 37. Dependence of the overall pressure in a circulation system of CO_2 and H_2 on the reaction time [145]

Table 9. Products from plasma reduced CO_2 and H_2 passed over a Co—Zr[a] catalyst [145]

Reaction mixture	Product gases, %				Conversion, %	
$CO_2:H_2$	CO_2	H_2	CO	CH_4	$CO \rightarrow CO_2$	$CO_2 \rightarrow CH_4$
1:3	12.5	0	4.4	83.1	9.4	62
1:3.29	16.1	6.3	0	77.6	13	63
1:3.3	12.5	10.5	0	77	10	62

[a] $Co:ZrO_2:MgO$:kieselguhr = 100:20:6:200

reconversion of CO into CO_2 was also observed, probably as the result of the reaction (see Sect. 5.1.1):

$$2\,CO \rightarrow CO_2 + C \tag{26}$$

It was necessary to choose the composition of the initial $CO_2:H_2$ mixture (Table 9) such that upon completion of the plasma reaction a mixture is obtained with the ratio $CO:H_2 = 1:2$, which is optimum for the synthesis of organic compounds in the catalytic stage.

When the heating of the Co—Zr catalyst was avoided [146] by diluting it with silica gel in the proportion of 1:1 appreciable amounts of organic substances, both liquid and solid, were formed. The main liquid products condensing at 195 K were hydrocarbons (Table 10) and water, the highest degree of conversion into these substances reported being 48.5%. The solids which were formed on the catalyst and silica gel were paraffinic hydrocarbons with melting points higher than 190–200 °C. The

Table 10. Condensed products from the plasma reaction of CO_2 and H_2 in a circulation system with catalysts placed in the afterglow [145–148]

Catalyst composition	Catalytic reaction started upon completion of the plasma reaction	Plasma and catalytic reactions proceed at the same time sequentially
Co—Zr catalyst ($Co:ZrO_2:MgO$:kieselguhr = 100:20:6:200)	Ethanol, formaldehyde, hydrocarbons (not analyzed)	—
(Co—Zr):silica gel = 1:1	Heptane, octane, isooctane, nonane, decane, isodecane, undecane	—
Ammonia catalyst ($Fe_2O_3:FeO:Al_2O_3:CaO:K_2O$ = 60.57:33.24:2.87:2.58:0.74%)	2,2-dimethylpentane, heptane, 2,4-dimethylhexane, 2,3-dimethylhexane, octane, 2,5-dimethylpentane, nonane, decane, undecane	acetaldehyde, methanol, ethanol, hexane, octane, isooctane, nonane, isononane, decane, isodecane, undecane, dimethylhexane
Methanol catalyst ($CuO:ZnO:Al_2O_3$ = = 2:1:0.16)	methanol, carbon black	methanol, carbon black

degree of conversion into solid paraffins was 11.9–19.6% for those remaining on the catalyst and 11.9–21.5% for those remaining on silica gel. The catalyst could be regenerated by periodic distillation of these substances in an atmosphere of H_2 at 550–600 °C. There was no elemental carbon on the catalyst and silica gel.

Other catalysts that have been used in these experiments [147,148] include an Fe (ammonia) catalyst and a Cu/Zn (methanol) catalyst. Table 10 gives the products condensed at 195 K from the catalytic reaction started upon complete reduction of the $CO_2 + H_2$ mixture in the plasma (two-stage experiments) and from both plasma and catalytic reactions proceeding at the same time sequentially in a circulation system (one-stage experiments).

Only when the Fe catalyst was initially exposed to a $CO:H_2$ mixture in molar proportions CH_4 was formed and CO_2, reduced previously to CO in the plasma, was regenerated:

$$2\,CO + 2\,H_2 = CH_4 + CO_2 \qquad (27)$$

Calculations of the equilibrium concentrations of the reactants in reaction (27) at different temperatures and the experimental data have shown that in the temperature range 180–360 °C the $CO + H_2$ mixture should be converted almost entirely into CH_4 and CO_2. With the Cu/Zn catalyst CH_4 was not found in the gaseous products which consisted essentially of unreacted CO_2 and H_2.

In the single stage experiments the degree of conversion of CO into CH_3OH and carbon black were 1.57 and 51.0%, respectively. In the two-stage experiments 3.7–5.0% of CO was converted to CH_4, 1.9–2.62% of CO into CH_3OH, 20.4–32.6% of CO into carbon black, and 4.02–6.38% of CO into CO_2. Calculations of the equilibrium compositions of the gaseous mixtures in the reaction

$$CO + H_2 = C(s) + H_2O(g) \qquad (28)$$

showed that at the temperature favoring the synthesis of methanol ($\simeq$270 °C) the $CO:H_2$ mixture can be converted almost completely into carbon black. In the plasma-catalytic experiments at $\simeq$1 atm the reduction of CO to carbon black and the formation of methane compete with the synthesis of methanol as expected from thermodynamic considerations.

4.3 Catalytic Effects Occurring in the Cold Trap

Winkler and coworkers [149,150] found that the product yields from electrically dissociated hydrogen-oxygen systems depended upon arrangement and temperature of the cold trap. Both hydrogen peroxide and polyoxide yields increased with decreasing trap temperature. Coating the trap walls [139] with concentrated solutions of H_3PO_4 or KCl decreased the H_2O_2 yields considerably (inset, Fig. 35) indicating that reaction (24) was certainly not taking place at 77 K. In fact, only traces of H_2O_2 were found at $S/V > 8\ cm^{-1}$. However, when 1 N H_3PO_4 solution was used for coating the H_2O_2 yield was unaffected. More dilute solutions of H_3PO_4 increased the H_2O_2 yield as expected. Coating the trap with ice [139] increased the H_2O_2 yield

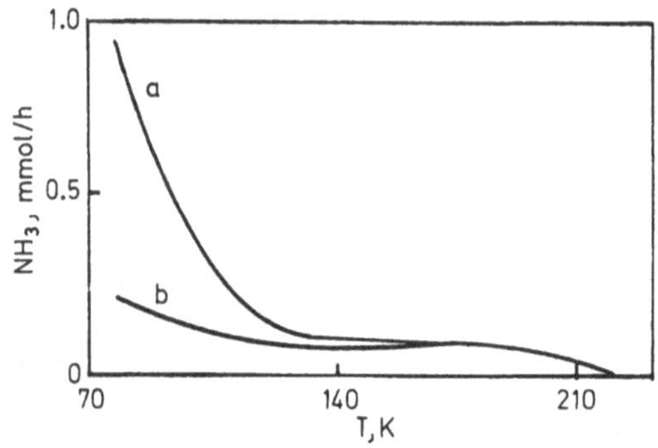

Fig. 38. Effect of trap temperature on NH_3 yield.
(a) Fe catalyst in plasma zone; (b) plasma gas passed over Fe catalyst [27)]

considerably, but the yield was independent of the relative thickness of the ice coating. Probably the ice-coated surfaces reduced the recombination of H atoms significantly and produced more O_2H which reacted with H to form more H_2O_2.

The yield of ammonia as a function of trap temperature [27)] in the plasma reaction of N_2 with H_2 is shown in Fig. 38. Below the trap temperature of 120 K the NH_3 yield increased indicating that ammonia was formed by reactions in the cold trap. When the trapping zone was covered with metal foils the NH_3 yields increased significantly [151)]. At 77 K, compared to Pyrex, Ag and Pt surfaces produced a relative increase in NH_3 yield of 1.23 and 1.35, respectively. If ammonia were formed by the reaction

$$NH + H_2 \rightarrow NH_3^* \tag{6}$$

then the energy removal is more effective on the metallic surface than on glass. Further work is necessary before these effects can be fully understood.

5 Catalytic Processes in Deposition and Etching of Solids

As in the previous sections, some of the processes to be discussed here do not obey the classical definition of catalysis. Nevertheless, they have a similar effect in promoting reactions which would, otherwise, not proceed at the given temperature because of a high kinetic barrier for thermal activation. In fact, some of the reactions would be even impossible to proceed thermally because of instability of the products at the high temperatures necessary for the thermal activation.

Electrons, ions, atoms, radicals and excited species are typically those promoting such reactions. In intense glow discharges, their concentrations reach significant values changing the overall reaction stoichiometry and, consequently, the steady-state concentrations of reactants and products. Thus, purely catalytic effects of the plasmas are observed only in weak discharges [1)].

The catalytic effects to be discussed include the processes taking place in the gas phase and on the surface of the solid. The former have the effect of promoting the gas-phase chemical reactions leading to the deposition at a significantly lower temperature than without the plasma. This is a principal advantage of low pressure plasma CVD (chemical vapor deposition) as compared to conventional methods, and it is being utilized in the fabrication of electronic devices, polymerization, preparation of protective coatings and others [46,152-154]. Processes taking place on the surface during deposition, such as ion bombardment and recombination of atoms and radicals influence the structural (and related) properties of the deposit. The final effect can be either beneficial (high density, stable films with excellent electronic properties) or deleterious (high stress in the film, amorphization), depending on the particular conditions [155,156]. Ion and electron bombardment of solids in an appropriate gaseous environment can also lead to enhanced rate of etching, nitriding, hydriding, oxidation and others.

A topic of great potential application which will not be discussed in any detail here is the *plasma treatment of solid catalysts*, the reason being the very limited number of papers published so far. Nevertheless, it is the opinion of the authors that this field will undergo very promising development in the near future. In many instances the treatment of solid catalysts is being performed in order to facilitate the creation of active sites or their regeneration after poisoning. The highly reactive components of a plasma can be utilized to perform such transformations at low temperatures thus avoiding, e.g. the problems of sintering and decrease of active specific area. Since, as a rule, only the surface of the catalysts is to be modified, there are apparently no limitations due to the low pressure at which glow discharges typically operate.

As in the whole field of catalysis, the crucial problem which arises is the incomplete mechanistic understanding of the processes under consideration. This is even more serious in the case of heterogeneous systems under plasma conditions.

5.1 Chemical Vapor Deposition

The subject involves essentially two different processes, the enhancement of the deposition rate and the modification of the properties of the deposit. Since the former has been discussed recently [1], emphasis will be on the second subject.

5.1.1 Control of the Deposition Rate

In many systems that are conventionally used for CVD, the chemical equilibrium is on the solid side, but high temperatures are required in order to achieve sufficient reaction rates. Several examples of such systems in use for fabrication of optical fibers [157] and electronic devices [1,152-154] are:

$$SiCl_4(g) + O_2(g) \rightarrow SiO_2(s) + 2\,Cl_2tg) \tag{29}$$

$$GeCl_2(g) + O_2(g) \rightarrow GeO_2(s) + 2\,Cl_2(g) \tag{30}$$

$$SiH_4(g) \rightarrow Si(s) + 2\,H_2(g) \tag{31}$$

$$3\,SiH_4(g) + 4\,NH_3(g) \rightarrow Si_3N_4(s) + 12\,H_2(g) \tag{32}$$

$$B_2H_6(g) \rightarrow 2\,B(s) + 3\,H_2(g) \tag{33}$$

In a glow discharge, the deposition can be facilitated with a high rate even at room temperature. However, impurities originating from the incompletely decomposed reactants, such as chlorine in SiO_2 and GeO_2 and hydrogen in Si, Si_3N_4, B, which are incorporated in large quantities at a too low deposition temperature degrade the properties of the deposit. Thus, higher temperatures have to be used (e.g. $\simeq 1000$ °C for optical fibers, 250 °C for a-Si, SiO_2, $\geqq 300$ °C for Si_3N_4 and $\geqq 400$ °C for B). In all these cases the deposition temperatures is less than that required in a conventional process. The advantage of using plasma to catalyze the reaction is the possibility to optimize the deposition temperature solely with respect to the materials' properties.

Very limited mechanistic studies of these reactions indicate that electron impact induced processes, such as dissociation and ionization, are the primary steps [158)]. This conclusion is likely to apply to most systems since the contribution of uncharged radicals to the overall reaction rates is small in weak discharges at low temperatures [1)]. However, caution is advisable in considering any mechanistic interpretations of such complicated processes. For example, more recent results show that ion- and electron-impact induced processes taking place on the surface of the deposit are those actually controlling the deposition rate of microcrystalline silicon [155,159,160)]. In their work Vepřek et al. have shown that, whereas increasing ion bombardment under a strongly negative bias increases the deposition rate only slightly, a dramatic decrease of the deposition rate is observed when the substrate potential is driven above the "wall potential" and it reaches zero value at a substrate potential which is still somewhat negative with respect to the surrounding plasma [159,160)].

As mentioned above, weak discharge operating at a low current can catalyze reactions which cannot be promoted thermally because of the instability of the reaction products at high temperatures. This applies to systems involving stable, strongly bonded educt molecules, such as N_2 and CO_2 and products which are unstable at high temperatures (e.g. MoN, WN). Example of such a system which has been studied in some detail is the Boudouard's reaction

$$2\,CO \rightleftarrows CO_2 + C(s)$$

Below $\simeq 300$ °C the equilibrium is far on the right-hand-side but the decomposition of carbon monoxide with formation of CO_2 and deposition of solid carbon proceeds only at an extremely slow rate because of the high dissociation energy of CO [161,162)]. A weak discharge operating at a current density of a few mA/cm^2 efficiently enhances the reaction rate [163)], the probable reaction mechanism being the dissociative attachment on CO followed by associative detachment of O^- and CO [1)].

Weak discharges are also being used in order to promote polymerization of hydrocarbons and halocarbons. The plasma polymerization proceeds via a chain reaction, initiation of which occurs by electron impact induced formation of free radicals. This subject has recently been summarized in several excellent review articles [152,164–166)] and the reader is referred to them for further information.

5.1.2 Control of the Structural Properties of the Deposit

The control of the structural and, consequently also of chemical and physical properties of the deposit can be achieved via discharge induced modification of the gas phase composition, or via particle-impact-induced surface processes, or a combined effect of both. In the following section we shall discuss several illustrative examples.

5.1.2.1 Discharge-Induced Modification of the Gas Phase Composition

In many instances the evaporation and deposition of solids is kinetically hindered. This applies particularly to associative and dissociative sublimation in which case evaporation and condensation coefficients down to 10^{-6} were found [167-169]. Examples are:

Dissociative sublimation

$$MeN(s) \rightarrow Me(g) + {}^{1}/_{2}\, N_2(g) \quad (34)$$

with Me: Al, Ga, Mg, Ti, Si, . . .

$$CdS(s) \rightarrow Cd(g) + {}^{1}/_{2}\, S_2(g) \quad (35)$$

$$EuCh(s) \rightarrow Eu(g) + [(1-\alpha)/2]\, Ch_2(g) + \alpha Ch(g) \quad (36)$$

with Ch: S, Se, Te

Associative sublimation

$$AsO_{3/2}\ (\text{Claudetite}) \rightarrow {}^{1}/_{4}\, As_4O_6(g) \quad (37)$$

$$P\ (\text{black, red}) \rightarrow {}^{1}/_{4}\, P_4(g) \quad (38)$$

$$NaCl(s) \rightarrow (1-\alpha)\, NaCl(g) + (\alpha/2)\, (NaCl)_2(g) \quad (39)$$

Upon condensation, a high activation energy is necessary in order to dissociate the gaseous species, such as N_2, S_2, Ch_2, As_4O_6, P_4. If such an energy is not provided, nonstoichiometric deposits, such as MeN_{1-x}, CdS_{1-x}, $EuCh_{1-x}$ or metastable modifications, P_4 (s, white), As_4O_6 (Arsenolite), result. In some cases both the evaporation and condensation can be facilitated by proper catalysts. Some examples are [170-172]:

$$P(\text{s, red, black}) \xrightarrow{Tl(l)} P_4(g) \quad (40)$$

similarly: As

$$P(\text{s, red}) \overset{I(ads)}{\rightleftharpoons} P_4(g) \quad (41)$$

similarly: Se, Te

$$AsO_{3/2}(\text{Claudetite}) \xrightarrow{I(ads)} As_4O_6(g) \quad (42)$$

Instead of iodine, aluminium trichloride can be used to catalyze reactions (41) and (42) [172].

Except for reaction (36) which has not been studied under plasma conditions so far, it has been shown that in all the systems given above either the evaporation or the deposition, or both, can be accelerated by low pressure plasmas [1,169,173-177].

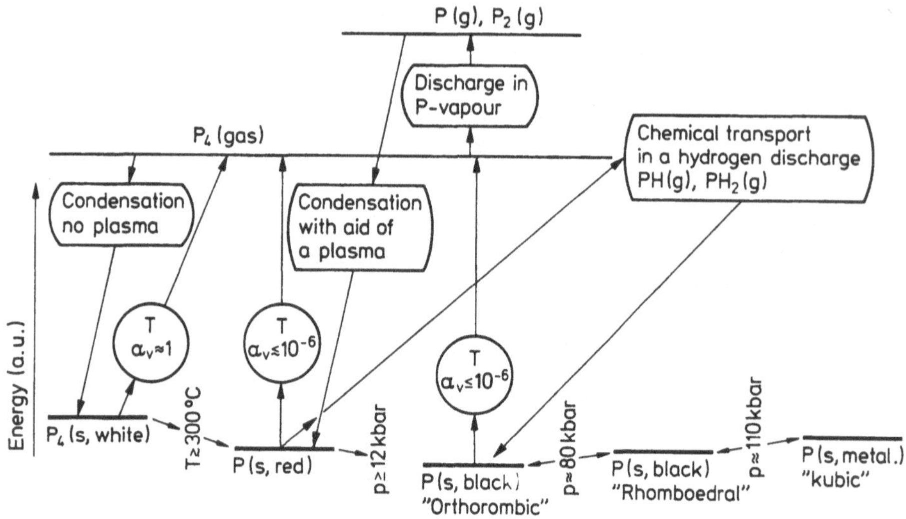

Fig. 39. Effect of glow discharge on the evaporation and deposition of phosphorus. T, thermal evaporation; α_v, evaporation coefficient

In particular, the deposition of phosphorus under catalytic effect of discharge plasmas was studied already in the last century (for summary of the literature see Refs. [176,178]) and it represents the most illustrative example to be discussed here.

The sublimation of phosphorus with and without the aid of discharge plasma is schematically shown in Fig. 39, together with the various solid polyforms and the conditions for their mutual transformations [179]). The thermodynamically stable black phosphorus as well as the slightly metastable red phosphorus evaporate associatively with formation of tetrahedral P_4 molecules. Upon nonactivated condensation these molecules remain preserved and the metastable, white phosphorus is deposited. Thus, both the evaporation as well as the condensation of red and black phosphorus is kinetically hindered. The effect of the discharge maintained in the phosphorus vapor is to dissociate these valence saturated P_4 molecules, which facilitates conditions for the formation of the polymeric red phosphorus. By means of chemical transport even the stable, orthorhombic modification, amorphous as well as polycrystalline has been obtained [1,155,177]).

A similar effect of the discharge has been demonstrated for the sublimation of arsenic oxide, As_2O_3. In this case the equilibrium vapor over solid As_2O_3 consists of As_4O_6-molecules, each being formed of four $AsO_{3/2}$ coordination polyhedra [180]). Upon non-activated condensation, the cubic, molecular crystal lattice of Arsenolite is formed even at temperature above −26 °C where it is metastable. If a glow discharge is maintained in the vapor, the As_4O_6-molecules are broken up into fragments and the Claudetite modification is deposited [174]).

The properties of the deposit, in particular its stoichiometry, can be influenced also in the case of dissociative sublimation if one of the gaseous species forms a relatively stable molecule. Nitrides, eq. (34), represent a classical example. In most of these systems, especially in the case of dielectrics on which dissociative chemisorption of nitrogen requires a high activation energy the partial pressure of atomic nitrogen is orders of magnitude less than that of N_2 and of the respective metal at equilibrium with the solid phase. Therefore, nitrogen-deficient solid, or even an almost nitrogen free metal (e.g. Al) is deposited upon non-activated sublimation. The dissociation of N_2 can be readily facilitated in a glow discharge [173] via vibrational-vibrational-pumping mechanism [7] and, as result, stoichiometric nitrides are formed. Besides the above-mentioned Si_3N_4:H, high quality dielectric films of P_3N_5 and $P_3N_5H_{x-1}$, $x \geqq 1.4$ [181], were deposited at $\leqq 350$ °C and single crystals of AlN, TiN [173] and of Si_3N_4 [182] were grown at a relatively low temperature of $\leqq 1100$ °C.

The activation of the gas phase enhances also the rate of the kind of reactions given by eq. (43)

$$A(s) + (m/n)\, B_n(g) \rightarrow AB_m(s) \qquad (43)$$

whenever the dissociative chemisorption of the gaseous molecules B_n on the AB_m-surface is the rate determining step without the discharge. Since high atom fluxes of the order of $\geqq 5 \times 10^{18}$ cm^{-2} s^{-1} impinging on the surface are obtained already in a weak discharge with $\geqq 1\%$ of dissociation at a total pressure of 1 torr, the rate determining step of the reaction becomes the bulk diffusion. Nitriding of metals is a typical example of such kind of reactions which were studied in some detail (e.g. Ref. [183-190]). Enhancement of the reaction rate by a glow discharge has also been observed for hydriding of metals [191,192] and it is expected to occur in far more heterogeneous systems than those studied so far.

5.1.2.2 Surface Processes Influencing the Properties of the Solid

Two cases are to be distinguished: Surface processes influencing the chemical reaction described by eq. (43) and those affecting the structural properties of the deposit. As in the previously discussed examples there is a more or less continuous transition between the catalytic effect of weak discharges and a "thermodynamic" effect of intense discharges depending on the extent of the plasma-surface-interaction and on the energy of the impinging species.

The rate of reaction (43) can be enhanced by ion- or electron-impact-induced dissociation of physisorbed molecules [193] having a similar effect as the activation in the gas phase (Sect. 2.1.2.1). Such effects probably occur also during the "ion-nitriding" of stainless steel but, in this case, the mechanism is complex and not well understood [183]. To the same category belong the recent studies of Rabalais and coworkers on the reactions of low and medium energy ions with solid surfaces [194-197]. Increasing the energy above a certain threshold leads to ion implantation and the interaction becomes more complex [198].

In considering the effect of surface processes on the structural, and related chemical and physical properties of the deposit a distinction has to be made between the effect of bombardment by energetic ions and chemical processes taking place on the surface under action of thermal or low energy species ($E_{kin} \leqq$ 5–15 eV). The former is observed if the deposition takes place on the cathode or under a large negative bias of the substrate whereas the latter dominates at floating potential which is only several volts negative with respect to the plasma. In spite of the large number of experimental papers the understanding of the processes under consideration is rather poor and any attempt at generalization and formulation of summarizing guidelines is a rather difficult and risky task. Nevertheless the present authors believe that such an attempt with all its limitations is more worthy than a descriptive review of the experimental work which was, in most cases, performed under different conditions and therefore does not allow a fair comparison.

In summary, the available experimental data suggest that exothermic chemical and physical processes on the surfaces and bombardment by low energy species ($E_{kin} \leqq 15$ eV) have the effect of facilitating the formation of thermodynamically more stable structural modifications whereas bomb;rdment by energetic ions ($E_{kin} \geqq 10^2$ eV) promotes the formation of metastable phases. Thus, only the former effect should be regarded as catalytic one. In the following, examples will be given to illustrate these general trends.

Among the most attractive tasks for a plasma chemist is the possibility of *preparation of diamond*, a high temperature and high pressure modification of carbon. The conventional synthesis of diamond is performed in its stability region at a pressure of several ten kilobars and several thousand degrees Kelvin, using suitable solvent catalysts in order to overcome the kinetic barrier for the transition from sp^2 to sp^3 hybridization of C-atoms. Without a catalyst the graphite → diamond transition occurs only at pressures of several hundred kilobars [199,200].

The preparation of diamond in the metastable region at a pressure of $\simeq 1$ bar can be performed *via* epitaxial growth on a diamond substrate from suitable gases such as CH_4 and CO containing C atoms promoted to hybrid states of high energy [201–203]. Following Ostwald's rule of stages [167], the metastable diamond phase can be quenched. Although of great scientific interest, this method is, however, of no practical use for industrial production.

From the practical point of view a more viable method of preparation of diamond and "diamond-like" carbon materials is the low pressure discharge assisted CVD and related methods, such as ion beam deposition. Among the number of papers available, there seems to be some confusion about the experimental conditions used and the results obtained.

Schmellenmeier [204a] was probably the first to report on the deposition of hard, carbonaceous films on the cathode of a dc discharge in gaseous hydrocarbons. He verified by means of X-ray diffraction that the deposit contained crystalline particles with cubic lattice showing two strong Bragg diffraction lines of diamond, the typical lines of graphite being absent [204b].

About twenty years after the pioneering work of Schmellenmeier a number of scientific papers appeared on this subject. These can be divided into two groups:

Aeisenberg and Chabot [202] and others [203, 205, 206] used carbon-ion beams with energies of ~50–100 eV to deposit hard, transparent films of high electric resistivity in which microcrystals with diamond lattice were identified by X-ray and electron diffraction methods. A related technique — sputtering of carbon target with a simultaneous bombardment of the substrate with low energetic ions (<100 eV) — has been used by Weissmantel and coworkers (see Ref. [207] and the references therein) for deposition of carbon films in which several high pressure metastable phases were identified together with the cubic diamond.

Most of the researchers have used cracking of hydrocarbons in an R.F. glow discharge for producing thin films of variable properties ranging from hydrocarbon polymers [164–166] to hard, amorphous, "diamond-like" films [209–214]. The properties of the films are strongly dependent on the deposition temperature, the nature of the hydrocarbon gas used (e.g. Ref. [214]) and on the bias of the substrate. A large negative bias and low deposition temperature favor the formation of the "diamond-like" films whereas graphitic carbon is obtained at a temperature above ~600 °C and floating potential [163]. The physical properties of the "diamond-like carbon" are intermediate between those of graphite and diamond and such a material contains significant quantities of incorporated hydrogen (10–25 at%) [1]. The electron spin resonance study by Gambino and Thompson [211] indicated that a significant quantity of the carbon atoms are in the graphitic sp^2 hybrid state. Although the understanding of the structural properties of this material is far from being comprehensive the available data favor a model of hydrogenated, disordered network with a variable mixing of the aromatic, sp^2, and diamond-like, sp^3, carbon atoms. Thus, in agreement with the general rule outline above, intense bombardment by energetic atoms favors the formation of metastable phases.

The discharge activated deposition of *amorphous and microcrystalline silicon* (a-Si and μc-Si, respectively) is presently among the most important applications of plasma chemistry [215]. This field has been opened up by Spear and Le Comber who where the first to demonstrate in 1975–76 the possibility of substitutional doping of a-Si [216], but only recently some, although still incomplete, understanding of the chemical processes during the deposition has been achieved [155,156,158–160,217]. The catalytic effect of the plasma on the decomposition of silane, reaction (31), was already mentioned in Sect. 5.1.1. Significant data on the mechanism of the gas phase reaction were provided in particular by Turban et al. [158]. However, very limited data are available on the role of surface processes on the deposition of a-Si. In the case of the preparation of μc-Si by means of an intense discharge in a mixture of H_2 with ≦1 at% SiH_4 the deposition takes place only if the electric potential of the substrate is negative (floating or biased) with respect to the plasma [159,160]. At a floating potential in a dc discharge the deposition rate is directly proportional to the ion flux density to the substrate with a proportionality factor of ~0.7 [160]. The probable interpretation has been given in terms of ion-impact induced fragmentation of a physi-sorbed SiH_4 followed by chemisorption of the SiH_x fragment and subsequent decomposition into solid silicon and H_2 [155]. This result points to the possibility that such surface processes may control the deposition rate also in other system in which the gaseous monomers (stable in the plasma) display a low sticking probability for the chemisorption.

The most interesting aspect of this system is the possibility of controlling the structural properties of the silicon films *via* a proper control of the surface processes which allow to deposit, at the same temperature, either a-Si or μc-Si. Furthermore, it is also possible to control, within certain limits, the mean crystallite size of μc-Si by a proper choice of the substrate bias at, otherwise constant discharge parameters and deposition temperature.

The decisive parameter controlling whether a-Si or μc-Si will be formed is the departure of the system from a partial chemical equilibrium [159,217]. (We consider here the typical range of $T_d \simeq 70$–400 °C.) In a weak discharge in silane, which is far away from such an equilibrium, i.e. the residence time of the species in the plasma zone, t_{res}, is small compared with the characteristic time, τ, of reaction (31),

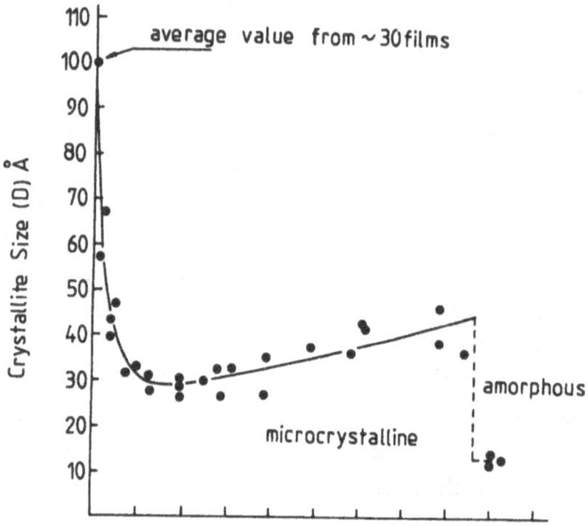

Fig. 40. The effect of substrate bias on the crystalline size of μc-Si (see text). At $V_b \leqq -900$ V a-Si is formed; the corresponding D-value should be considered only as a reference number illustrating the discontinuous nature of the transition. $T_d = 260$ °C, dc discharge current = 126 mA. diameter of discharge tube = 5 cm, p = 0.37 mbar

a-Si is deposited whereas μc-Si is obtained in an intense discharge when the system approaches partial chemical equilibrium, i.e. $\tau \ll t_{res}$. The crystalline-to-amorphous transition has been studied recently in some detail [218]. Accordingly, the formation of the thermodynamically more stable μc-Si phase is catalyzed by exothermic recombination processes taking place on the surface of the growing film.

In agreement with the general rule outlined above, bombardment of the growing film by energetic ions results in the formation of a less stable phase, either μc-Si of a smaller crystallite size or even a-Si. This is illustrated in Fig. 40 which shows the dependence of the crystallite size on the bias and the discontinuous transition to a-Si at $V_b \leqq -900$ V (see Ref. [219] for further details). It is seen that the ion bombardment causes a decrease of the crystallite size as compared to that obtained at the floating potential. The apparent increase of the crystallite size between $V_b \simeq -200$ and ~ -800 V is at the present time somewhat difficult to understand because of problems associated with separating the effect of the finite crystallite size and that of stress in the films on the broadening of the X-ray diffraction lines. At bias larger than ~ -900 V only amorphous films are obtained; this has been verified also by other methods as well [218,219]. Since with decreasing crystallite size the free energy of μc-Si increases as compared with that of a single crystal Si and a-Si is metastable even with respect to μc-Si of a crystallite size D ≧ 30 Å, [218] these results confirm in a consistent way the general rule that bombardment of the growing film by energetic ions causes the formation of less stable phases.

5.2 Chemical Evaporation of Solids

The chemical evaporation, also called "etching", eq. (44):

$$aA(s) + (b/n)\,B_n(g) \rightarrow A_aB_b(g) \tag{44}$$

can be catalyzed either via formation of reactive radicals (e.g. atoms) in the gas phase or by bombardment of the solid surface with ions or electrons of a medium energy ($\leqq 100$ eV). In the former case the activation energy is simply provided by the dissociation of the stable, gaseous monomers and the resulting fragments, typically atoms, subsequently react with the solid. The effective activation energy as obtained from the temperature dependence of the reaction rate [eq. (44)] amounts typically to a few kcal/mole or it can be even negative indicating a complex mechanism of the surface reaction with some intermediate, exothermic pseudo-equilibrium of the chemisorbed species. Included in the many systems that have been studied are the reactions of atomic oxygen with carbon, molybdenum, tungsten, atomic chlorine and fluorine with boron, carbon, molybdenum and tungsten [220)] and atomic hydrogen with carbon [221)].

The reactive species can also be produced directly on the surface by ion- or electron-impact induced fragmentation of the physisorbed molecules of the reactant B_n (eq. 44). An example of this kind is the etching of Si, SiO_2 and other materials by XeF_2 upon bombardment of the surface by Ar^+ ions or electrons. Alternatively, the molecular ion B_n^+ can dissociate upon collision with the surface and, subsequently, undergo chemical reaction. This kind of processes have been reviewed recently by Winters [222)]. At ion energies above the threshold value for sputtering the situation is more complex and physical and chemical sputtering [198)] usually dominates.

The mechanistic understanding of such surface processes is in general very limited. Most of the data available were obtained in experiments with atom-, ion-, and electron-beams under conditions where the particle fluxes are orders of magnitude less than those in the glow discharge. Since the typical surface processes have complex mechanisms with parallel reaction paths, scaling of the results of the beam experiments to the plasmas at 10^{-2} — 1 torr is difficult and in general not justified. Nevertheless, they represent a very significant contribution which is highly desirable for a better understanding of the plasma chemistry. The reactive ion etching of silicon (and other materials) is probably the best example illustrating this point. This field has been recently summarized in several excellent review papers [152,166,222–224)] and, therefore, it will not be discussed here.

As a result of numerous studies a basic understanding of the mechanism of electron impact induced desorption was achieved in the past decade [222,225–228)]. In contrast, few papers have been published on the electron impact induced chemical evaporation. Besides the above-mentioned etching of Si and some other materials by XeF_2 [222)], only the C/H_2-, TiC/H_2-, TiB_2/H_2- [229)] and the Si/H_2-systems [230)] have been investigated. In the case of silicon etching by hydrogen the effect of the electron impact cannot be considered as a purely catalytic one since the reaction product, silane, is thermodynamically unstable with respect to solid silicon and H_2. Nevertheless, it provides some insight into the mechanism of such processes and, therefore, we shall discuss it briefly.

The reaction mechanism is summarized, in eq. (45):

$$(\alpha/2)H_2(g) \underset{k_{-1}}{\overset{k_1}{\rightleftharpoons}} (\alpha/2)H_2(ads) \underset{k_{-2}}{\overset{k_2}{\rightleftharpoons}} \alpha H(ads) \rightarrow \cdots\overset{k_r}{\cdots} \rightarrow SiH_x(g) \tag{45}$$

where $x = 3, 4$ and $\alpha = 1$ or 2. There are two crucial points to be emphasized: The electron bombardment has the effect of promoting the dissociation of the physisorbed molecular hydrogen, k_2, and the process competing with silane formation is repre-

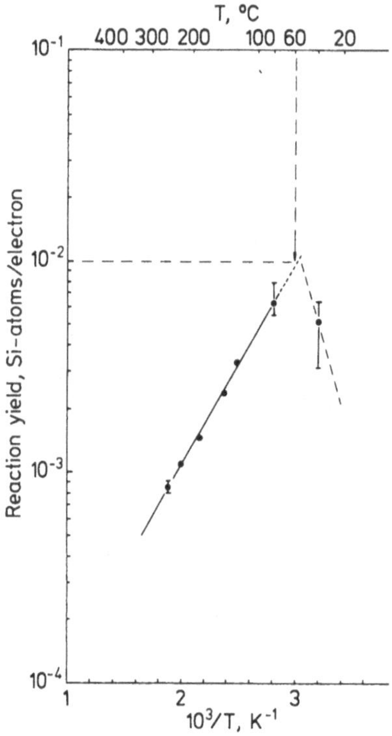

Fig. 41. Arrhenius plot of reaction yield for electron impact induced etching of x-Si by hydrogen

sented by the thermally activated recombination, k_{-2}. With increasing temperature the recombination leads to depletion of the chemisorbed hydrogen, H(ads) and, consequently, the overall reaction rate (i.e. the etching rate) displays a negative activation energy as shown in Fig. 41. Because of the strong covalent bonding of the surface SiH_x ($x \leqq 3$) groups to the bulk silicon, the cross section for electron impact induced desorption of the chemisorbed hydrogen is expected to be very low as compared with that for dissociation of H_2(ads). Therefore, the electron impact enhances the etch rate.

The generally observed decrease of the cross section for electron stimulated desorption with increasing bond strength between adsorbate and substrate suggests that such a selective enhancement of the etch rate is to be observed in any solid/gas system in which the dissociative chemisorption of the gaseous molecules on the surface requires a high activation energy and the energy of chemisorption is sufficiently high.

6 Concluding Remarks

Looking back over the development of plasma catalysis the authors note that several plasma systems have been studied in the presence of different materials that are known to be catalysts in chemical reactions. However, most studies have been of an empirical rather than fundamental nature. In consequence the current knowledge of catalytic-reaction paths in plasmas is quite tenuous. This is to be attributed mainly to a lack of understanding of the surface processes in plasma catalysis. In preparing this review the authors were motivated by the need for a new impetus in this field rather than by its maturity.

Studies of catalysts used in conventional chemical reactions suggest that it is not entirely defensible to correlate the properties of a catalyst, which are basically concerned with its surface, with properties such as work function and, in the case of transition metals, with d-bond character. In chemisorption and in catalysis, the chemical properties of the individual surface species are more important than the collective properties of metal catalysts such as can be deduced from the band model. The most significant result is the correlation found between a low heat of chemisorption for the adsorbate and a high catalytic activity for the adsorbant.

Progress may be achieved in understanding the phenomena of chemisorption in plasma catalysis, as it has indeed in the field of conventional catalysis, by following a catalytic reaction on a given catalyst surface through observation of the adsorbed phases and analysis of the desorbed phases as a function of the various parameters, namely, temperature and composition of the plasma. Low-energy electron diffraction, Auger spectroscopy, and the modern methods of analysis of residual gases may be adapted to this kind of study.

With respect to industrial application the catalytic effects of plasmas during CVD and etching of solid materials are presently the most important. Here again, further work towards a better mechanistic understanding of the process is highly desirable.

In gas phase synthesis under plasma conditions the use of catalysts is probably the only way to improve the selectivity and reaction yields. Very little has been done so far on the application of low pressure plasmas for treatment of solid catalysts. The authors hope that both of these problems will be solved on the basis of research of a very fundamental nature.

7 References

1. Vepřek, S.: Chimia *34*, 489 (1980); Pure Appl. Chem. *48*, 163 (1976)
2. Venugopalan, M.: Internat. Workshop on Plasma Chemistry in Technology, Ashkelon, Israel, Apr. 1981
3. Rapakoulias, D., Gicquel, A., Amouroux, J.: Proc. Fifth Inter. Symp. Plasma Chem., Edinburgh, Aug. 1981, p. 688
4. Eremin, E. N.: Russ. J. Phys. Chem. *49*, 1112 (1975)
5. See papers in Proc. Inter. Symp. Ozone and Water, Berlin, 1977. Berlin: O. Hess 1978
6. See papers in the Specialized Session on Ozone Synthesis in Proc. Fifth Inter. Symp. Plasma Chem., Edinburgh, Aug. 1981, pp. 415–471
7. Capitelli, M., Molinari, E.: In: Plasma Chemistry II (Topics Curr. Chem., *90*) (Vepřek, S., Venugopalan, M., ed.), p. 59. Berlin—Heidelberg—New York: Springer 1980
8. Venugopalan, M.: In: Reactions Under Plasma Conditions, Vol. II. (Venugopalan, M., ed.), New York: Wiley/Interscience, 1971
9. For a brief historical development of the work, see Emel'yanov, Yu. M., Gvozd', V. F., Bogorodskii, M. M.: Russ. J. Phys. Chem. *48*, 403 (1974)

10. Locqueneux, M., Etile, F., Ben-Aim, R. I.: Proc. Fourth Inter. Symp. Plasma Chem., Zurich, 1979, p. 366
11. Fominskii, L. P., Gorozhankin, E. V., Men'shov, V. N.: Russ. J. Phys. Chem. *51*, 541 (1977)
12. Komel'kov, V. S., Modzolevskii, V. I.: ibid. *52*, 889 (1978)
13. Venugopalan, M., Rajaei-Rizi, A.: Z. Physik. Chem., Neue Folge, *126*, 15 (1981)
14. Eremin, E. N., Mal'tsev, A. N.: Zh. Fiz. Khim. *30*, 1179 (1956)
15. Vasil'ev, S. S., Kobozev, N. I., Eremin, E. N.: ibid. *7*, 619 (1936)
16. Eremin, E. N.: ibid. *32*, 2543 (1958).
17. Pollo, I., Mal'tsev, A. N., Eremin, E. N.: Russ. J. Phys. Chem. *37*, 1130 (1963)
18. Mal'tsev, A. N., Eremin, E. N., Vorob'eva, I. N.: ibid. *33*, 79 (1959)
19. Mal'tsev, A. N., Eremin, E. N., Ivanter, V. L.: ibid. *39*, 1012 (1965)
20. Gvozd', V. F., Bogorodskii, M. M., Emel'yanov, Yu. M.: ibid. *48*, 1333 (1974)
21. Emel'yanov, Yu. M., Gvozd', V. F.: ibid. *52*, 68 (1978)
22. Gvozd', V. F., Emel'yanov, Yu. M.: ibid. *52*, 360 (1978)
23. Mal'tsev, A. N., Eremin, E. N., Borisova, S. I.: Vestnik Moskov. Univ., Khim. 2, 34 (1967)
24. Brewer, A. K., Westhaver, J. W.: J. Phys. Chem. *34*, 153 (1930)
25. Simonyan, L. M., Syaduk, V. L., Eremin, E. N.: Russ. J. Phys. Chem. *52*, 735 (1978)
26. Chapman, B.: Glow Discharge Processes. New York: Wiley-Interscience, 1980, p. 79
27. Botchway, G. Y., Venugopalan, M.: Z. Physik. Chem., Neue Folge *120*, 103 (1980)
28. Suyuki, M., Takahashi, S., Miyazoki, S.: J. Chem. Soc. Japan, Pure Chem. Sec. *75*, 1265 (1954); *76*, 32 (1955)
29. Devins, J. C., Burton, M.: J. Am. Chem. Soc. *76*, 2618 (1954)
30. Rathsack, H. A.: Z. Physik. Chem. *142*, 101 (1960)
31. Baskakov, I. B., Zimina, I. D., Maksimov, A. I., Svettsov, V. I.: Russ. J. Phys. Chem. *47*, 1339 (1973)
32. Sharp, T. E., Dowell, J. T.: J. Chem. Phys. *50*, 3024 (1969)
33. Svettsov, V. I., Maksimov, A. I., Zimina, I. D.: Russ. J. Phys. Chem. *49*, 801 (1975)
34. Simonyan, L. M., Syaduk, V. L., Eremin, E. N.: ibid. *52*, 258 (1978)
35. Syaduk, V. L., Eremin, E. N.: ibid. *49*, 396 (1975)
36. Carbaugh, D. C., Munno, F. J., Marchello, J. M.: J. Chem. Phys. *47*, 5211 (1967)
37. Hanes, M. H., Bair, E. J.: J. Chem. Phys. *38*, 672 (1963)
38. Thornton, J. D., Charlton, W. D., Spedding, P. L.: Adv. Chem. Ser. *80*, 165 (1969)
39. Ghosh, P. K., Bair, E. J.: J. Chem. Phys. *45*, 4738 (1966)
40. Cramarossa, F., Colaprico, V., d'Agostino, R., Ferraro, G., Molinari, E.: Proc. Third Inter. Symp. Plasma Chem., Limoges, 1977, p. G. 5.13
a) d'Agostino, R., Cramarossa, F., De Benedicto, S., Ferraro, G.: Plasma Chem. Plasma Processing, *1*, 19 (1981)
41. Syaduk, V. L., Eremin, E. N.: Russ. J. Phys. Chem. *49*, 310 (1975)
42. Syaduk, V. L., Eremin, E. N.: ibid. *52*, 396 (1978)
43. Syaduk, V. L., Eremin, E. N.: ibid. *49*, 536 (1975)
44. Syaduk, V. L., Eremin, E. N.: ibid. *49*, 538 (1975)
45. Foner, S. N., Hudson, R. L.: J. Chem. Phys. *45*, 40 (1966)
46. Drost, H.: Plasmachemie, Akademie Verlag, Berlin 1978, pp. 296–311
47. Rubtsova, E. A., Eremin, E. N.: Russ. J. Phys. Chem. *51*, 1737 (1977)
48. Rubtsova, E. A., Eremin, E. N.: ibid. *51*, 1739 (1977)
49. Weissler, G. L.: Phys. Rev. *63*, 96 (1943)
50. Das, M. K.: Z. angew. Phys. *13*, 410 (1961)
51. Coffin, F. E.: J. Chem. Phys. *30*, 593 (1959)
52. Shaw, T. M.: ibid. *30*, 1366 (1959); *31*, 1142 (1959)
53. Zaitsev, V. V., Maksimov, A. I., Svettsov, V. I.: Russ. J. Phys. Chem. *47*, 761 (1973)
54. Polak, L. S.: Khim. Vys. Energ. *6*, 266 (1972)
55. Zaitsev, V. V.: Russ. J. Phys. Chem. *51*, 316 (1977)
56. Fillipov, Yu. V., Vendillo, V. P.: ibid. *35*, 303 (1961)
57. Kovaliv, B., Briner, E.: Helv. Chim. Acta *36*, 275 (1953)
58. Fillipov, Yu. V., Vendillo, V. P.: Russ. J. Phys. Chem. *36*, 1069 (1962)
59. Popovich, M. P., Zhitnev, Yu. N., Fillipov, Yu, V.: ibid. *45*, 123 (1973).

60. Donohoe, K. G., Shair, F. H., Wulf, O. R.: Paper at the Sympos. Plasma Chem. Processing, Atlantic City, New Jersey, Aug. 1976
61. Review: Knipovich, O. M., Emel'yanov, Yu. M., Fillipov, Yu. V.: Russ. J. Phys. Chem. *47*, 1474 (1974)
62. Zimina, I. D., Ilyushchenko, N. I., Maksimov, A. I., Svettsov, V. I.: ibid. *49*, 864 (1975)
63. Zimina, I. D., Maksimov, A. I., Svettsov, V. I.: ibid. *49*, 864 (1975)
64. Brooks, B. W., Spearot, R. M.: J. Appl. Chem. Biotechnol. *24*, 621 (1974)
65. Venugopalan, M., Jones, R. A.: Chemistry of Dissociated Water Vapor and Related Systems, Wiley-Interscience, New York, 1968; Chem. Rev. *66*, 133 (1966)
66. Roychowdhury, S., Cham, K., Roychowdhury, U. K., Venugopalan, M.: Paper at the Fourteenth Great Lakes Regional Meet. Amer. Chem. Soc., Macomb, IL., 1980
67. Roychowdhury, S., Roychowdhury, U. K., Venugopalan, M.: Unpublished (1981)
68. Brooks, B. W., Sambrook, R. M.: J. Appl. Chem. *22*, 9 (1972)
69. Brooks, B. W., Sambrook, R. M.: ibid. *22*, 839 (1972)
70. Semiokhin, I. A., Andreev, Yu. P., Panchenkov, G. M.: Vestnik Moskov Univ., Ser. Khim. *19* (5), 40 (1964); *20*(6), 24(1965)
71. Andreev, Yu. P., Semiokhin, I. A., Panchenkov, G. M., Utirov, B. U.: Russ. J. Phys. Chem. *40*, 1156 (1966)
72. Andreev, Yu. P., Voronkov, Yu. M., Semiokhin, I. A.: ibid. *49*, 394 (1975)
73. Blaustein, B. D., Fu, Y. C.: Adv. Chem. Ser. *80*, 259 (1969)
74. Blaustein, B. D., Steiner, W. A., Fu, Y. C.: J. Phys. Chem. *76*, 2941 (1972)
75. Drost, H., Mach, R., Klotz, H.-D.: Proc. Fourth Inter Symp. Plasma Chem., Zurich, 1979, p. 700
76. Meshkova, G. I., Eremin, E. N.: Russ. J. Phys. Chem. *44*, 255 (1970)
77. Bebeshko, G. I., Eremin, E. N.: ibid. *48*, 50 (1974)
78. Drost, H., Klotz, H.-D.: Proc. Third Inter. Symp. Plasma Chem. Limoges, 1977
79. Luk'yanov, V. B., Eremeev, A. P., Nesmeyanov, A. N.: Russ. J. Phys. Chem. *48*, 531 (1974)
80. Lebedev, V. P., Fillipov, Yu. V., Kobozev, N. I.: Zh. Fiz. Khim. *24*, 845 (1950)
81. Fillipov, Yu. V., Lebedev, V. P., Zalaman, V. V., Kobozev, N. I.: ibid. *24*, 1009 (1950)
82. Warburg, E., Leithaeuser, G.: Ann. Physik *28*, 17 (1908)
83. Starke, A.: Z. Elektrochem. *29*, 358 (1923)
84. Brit. Pat. 1265669 and 1265670 (1972); 1401692 (1975)
85. French Pat. 2182244 (1974)
86. Voblikova, V. A., Fillipov, Yu. V., Vendillo, V. P.: Russ. J. Phys. Chem. *54*, 1416 (1980)
87. Pollo, I., Jaroszynska-Wolinska, J., Ozonek, J.: Proc. Fifth Inter. Symp. Plasma Chem., Edinburgh, Aug. 1981, p. 445
88. Salge, J., Labrenz, M., Scheibe, K.: ibid. p. 427
89. Eliasson, B., Kogelschatz, U.: Proc. Fourth Inter. Symp. Plasma Chem., Zurich, Aug.–Sept. 1979, p. 729
90. Semiokhin, I. A., Pitskhelauri, E. N., Kobozev, N. I., Sindyukov, V. G.: Russ. J. Phys. Chem. *35*, 1301 (1961)
91. Rubtsova, E. A., Eremin. E. N., Mal'tsev, A. N.: ibid. *40*, 1671 (1966)
92. Rubtsova, E. A., Eremin, E. N.: ibid. *42*, 1099 (1968); *45*, 845 (1971)
93. Rubtsova, E. A., Eremin, E. N.: ibid. *45*, 1249 (1971)
94. Eremin. E. N., Mal'tsev, A. N., Syaduk, V. L.: ibid. *45*, 635 (1971)
95. Eremin, E. N., Mal'tsev, A. N., Belova, V. M.: ibid. *43*, 443 (1969); *45*, 205 (1971)
96. Eremin, E. N., Mal'tsev, A. N., Belova, V. M.: Doklady Akad. Nauk SSR *190*, 629 (1970)
97. Syaduk, V. L., Eremin, E. N.: Russ. J. Phys. Chem. *47*, 136 (1973)
98. Eremin, E. N., Mal'tsev, A. N., Rusakova, L. A.: ibid. *48*, 1229 (1974)
99. Eremin, E. N., Mal'tsev, A. N., Rusakova, L. A.: ibid. *48*, 1256 (1974)
100. Meshkova, G. I., Eremin, E. N.: ibid. *44*, 1294 (1970)
101. Rusakova, L. A., Eremin, E. N.: ibid. *49*, 439, 604 (1975)
102. Morinaga, K., Suzuki, M.: Bull. Chem. Soc. Japan, *35*, 429 (1962)
103. Morinaga, K.: ibid. *35*, 626 (1962)
104. Brunet, P. E., Deglise, X., Giguere, P. A.: Can. J. Chem. *48*, 2042 (1970)
105. Giguere, P. A.: ICSU Rev. *4*, 2989 (1965); J. Chem. Phys. *42*, 2989 (1965)
106. Hata, N., Giguere, P. A.: Can. J. Chem. *44*, 869 (1966)

107. Jones, R. A., Venugopalan, M.: ibid. *45*, 2452 (1967)
108. Jones, R. A., Chan, W., Venugopalan, M.: J. Chem. Phys. *51*, 1273 (1969)
109. Syaduk, V. L., Eremin, E. N.: Russ. J. Phys. Chem. *46*, 1644 (1972)
110. Syaduk, V. L., Eremin, E. N.: ibid. *47*, 864 (1973)
111. Syaduk, V. L., Eremin, E. N.: ibid. *49*, 587 (1975)
112. Syaduk, V. L., Eremin, E. N.: ibid. *47*, 137 (1973)
113. Mal'tsev, A. N., Eremin, E. N., Belova, V. M.: ibid. *45*, 1042 (1971)
114. Belova, V. M., Eremin, E. N., Mal'tsev, A. N.: ibid. *52*, 968 (1978)
115. Eremin, E. N., Belova, V. M., Mal'tsev, A. N.: ibid. *52*, 970 (1978)
116. Cavvadias, S., Rapakoulias, D., Amouroux, J.: Proc. Fourth Inter. Symp. Plasma Chem., Zurich, 1979, p. 378
117. Rapakoulias, D., Cavvadias, S., Amouroux, J.: Rev. Phys. Appl. *15* (7), 1261 (1980)
118. Eremin, E. N., Rubtsova, E. A.: Russ. J. Phys. Chem. *48*, 689 (1974)
119. For a review of early work see Bronfin, B. R.: Adv. Chem. Ser. *80*, 423 (1969)
120. Gartaganis, P. A., Winkler, C. A.: Can. J. Chem. *34*, 1457 (1956)
121. Capezzuto, P., Cramarossa, F., Giovanni, F., Maione, P., Molinari, E.: Gazz. Chim. Ital. *103*, 1153 (1973)
122. Capezzuto, P., Cramarossa, F., d'Agostino, R., Molinari, E.: Gazz. Chim. Ital. *103*, 1169 (1973)
123. a Rapakoulias, D., Amouroux, J.: Proc. Fourth Inter. Symp. Plasma Chem., Zurich, 1979, p. 372; b Rev. Phys. Appl. *15*, 1251 (1980)
124. Simionescu, Cr., Dumitriu, S., Bulacovschi, V., Onac, D.: Z. Naturforsch. *30b*, 516 (1975)
125. Schoetzau, H. J., Vepřek, S.: Appl. Phys. *7*, 271 (1975)
126. Eremin, E. N., Mal'tsev, A. N., Ivanter, V. L.: Russ. J. Phys. Chem. *52*, 204 (1978)
127. Eremin, E. N., Mal'tsev, A. N., Ivanter, V. L.: ibid. *54*, 614 (1980)
128. Eremin, E. N., Mal'tsev, A. N., Ivanter, V. L.: ibid. *54*, 80 (1980)
129. Rubtsova, E. A., Eremin, E. N.: ibid. *42*, 536 (1968)
130. Rubtsova, E. A., Eremin, E. N.: ibid. *47*, 35 (1973)
131. Mal'tsev, A. N., Churina, L. A., Eremin, E. N.: ibid. *42*, 1235 (1968)
132. Venugopalan, M., Botchway, G. Y.: Unpublished (1980)
133. Sabadil, H., Kastelewicz, H., Bachmann, P.: Proc. Fifth Inter. Symp. Plasma Chem., Edinburgh, 1981, p. 456
134. Jones, R. A., Chan, W., Venugopalan, M.: J. Phys. Chem. *73*, 3693 (1969)
135. Barton, S. S., Jones, R. A.: Chem. Commun. *17*, 406 (1965)
136. Venugopalan, M., Choi, K.: Z. Phys. Chem., Neue Folge *74*, 32 (1971)
137. Venugopalan, M., Shih, A.-L.: Plasma Chem. Plasma Proc., *1*, 191 (1981)
138. Jones, R. A., Barton, S. S.: Can. J. Chem. *41*, 1045 (1963)
139. Choi, K., Cook, J. E., Venugopalan, M.: Z. anorg. allg. Chem. *384*, 287 (1971)
140. Venugopalan, M., Botchway, G. Y., Suda, R. F.: Proc. Fifth Inter. Symp. Plasma Chem., Edinburgh, 1981, p. 694
141. Taniguchi, Y., Akashi, K.: J. Electrochem. Soc. Japan *46*, 560 (1978); Proc. Fourth Inter. Symp. Plasma Chem., Zurich, 1979, p. 494
142. Venugopalan, M., Nadizadeh, H.: Oxidation Communications, in press
143. Kálmán, J., Emódy, Z., Dudás, É.: Proc. Third Inter. Symp. Plasma Chem., Limoges, 1977
144. Kálmán, J., Izsáki, Z., Varga, T. A., Siklôs, P.: Proc. Fifth Inter. Symp. Plasma Chem., Edinburgh, 1981, p. 710
145. Eremin, E. N., Mal'tsev, A. N., Ivanter, V. L., Belova, V. M.: Russ. J. Phys. Chem. *52*, 1654 (1978)
146. Eremin, E. N., Mal'tsev, A. N., Belova, V. M., Ivanter, V. L.: ibid. *53*, 1450 (1979)
147. Eremin, E. N., Mal'tsev, A. N., Belova, V. M., Ivanter, V. L.: ibid. *53*, 1159 (1979)
148. Eremin, E. N., Mal'tsev, A. N., Belova, V. M., Ivanter, V. L.: ibid. *53*, 1448 (1979)
149. Jones, R. A., Winkler, C. A.: Can. J. Chem. *29*, 1013 (1951)
150. Batzold, J. S., Luner, C., Winkler, C. A.: ibid. *31*, 264 (1953)
151. Venugopalan, M., Yin, K. S.: Unpublished (1981)
152. Techniques and Applications of Plasma Chemistry (Hollahan, J. R., Bell, A. T., ed.), Wiley Interscience, New York 1974

153. Present Status and Future Trends in Plasma Chemistry and Arc Technology, NSF Workshop (Oskam, H., Pfender, E., ed.), Univ. Minnesota, Minneapolis, 1980
154. Thin Film Processes (Vossen, J. L., Kern, W., ed.), Academic Press, New York, 1978
155. Vepřek, S.: Pure Appl. Chem. 1982, in press
156. Sarott, F.-A., Iqbal, Z., Vepřek, S.: Solid St. Commun., 1982, in press
157. Kuppers, D., Lydtin, H.: In: Plasma Chemistry I (Topics Curr. Chem. *89*) (Vepřek, S., Venugopalan, M., ed.), Springer: Berlin—Heidelberg—New York, 1980
158. Turban, G., Catherine, Y., Grolleau, B.: Thin Solid Films *67*, 309 (1980), *77*, 287 (1981); Plasma Chem. Plasma Proc. *2*, 81 (1982)
159. Vepřek, S., Iqbal, Z., Oswald, H. R., Sarott, F.-A., Wagner, J. J.: J. Physique *42*, 251 (1981)
160. Sarott, F.-A.: Diplomarbeit, Univ. Zurich, 1981
161. Knippenberg, W. F., Lersmacher, B., Lydtin, H., Moore, A. W.: Philips Tech. Rev. *28*, 231 (1967)
162. Schafer, H.: Chemical Transport Reactions, Wiley, New York, 1964
163. Vepřek, S.: J. Crystal Growth *17*, 101 (1972)
164. Plasma Chemistry of Polymers (Shen, M., ed.), Marcel Dekker, New York—Basel, 1976
165. Bell, A. T.: In: Plasma Chemistry III (Topics in Curr. Chem. *94*) (Vepřek, S., Venugopalan, M., ed.), Springer, Berlin—Heidelberg—New York, 1980
166. Kay, E., Coburn, J., Dilks, A.: ibid.
167. Hirth, J. P., Pound, G. M.: Condensation and Evaporation, Pergamon Press, London, 1963
168. Somorjai, G. A., Lester, J. E.: In: Progr. Solid State Chem., *4* (Reiss, H., ed.), Plenum Press, London—New York, 1974
170. Brewer, L., Kane, J. S.: J. Phys. Chem. *59*, 105 (1955)
171. Schafer, H., Trenkel, M.: Z. anorg. allg. Chem. *391*, 11 (1972); *420*, 261 (1976)
172. Schafer, H., Binnewies, M.: ibid. *441*, 216 (1978)
173. Vepřek, S., Brendel, C., Schafer, H.: J. Crystal Growth *9*, 266 (1971)
174. Becker, K. A., Graf, G., Stranski, I. N.: Z. phys. chem., Neue Folge *30*, 76 (1961)
175. Becker, K. A., Niesler, R. A.: ibid. *36*, 53 (1963)
176. Gmelins Handb. Anorg. Chemie, 8th Edition, Vol. P/B. Verlag Chemie, Weinheim Bergstr., 1964
177. a. Vepřek, S., Oswald, H. R.: Z. anorg. allg. Chem. *415*, 190 (1975)
177. b. Wild, R.: Ph. D. Thesis, Univ. Zurich, 1979
178. Kohlschutter, V., Frumkin, A.: Z. Elektrochem. *20*, 110 (1914)
179. Krebs, H.: Grundzüge der anorganischen Kristallchemie, Ferdinand Enke, Stuttgart, 1968
180. Plieth, K., Stranski, I. N.: Z. Physik *156*, 360 (1959)
181. Vepřek, S., Iqbal, Z., Brunner, J., Scharli, M.: Phil. Mag. *B43*, 527 (1981)
182. Wirz, E., Vepřek, S.: Unpublished
183. Hudis, M.: In: Ref. 152, p. 113.
184. Konuma, M., Kanzaki, Y., Matsumoto, O.: Proc. Fourth Inter. Symp. Plasma Chem., 1979, p. 174
185. Matsumoto, O., Yatsuda, Y.: Proc. Fifth Inter. Symp. Plasma Chem., Edinburgh, Aug. 1981, p. 382
186. Akashi, K., Taniguchi, Y., Takada, R.: ibid. p. 376
187. Braganza, C., Stussi, H., Vepřek, S.: J. Nucl. Materials *87*, 331 (1976)
188. Braganza, C., Vepřek, S., Wirz, E., Stussi, H., Textor, M.: Proc. Fourth Inter. Symp. Plasma Chem., Zurich, Aug.-Sept. 1979, p. 100
189. Wirz, E., Oswald, H. R., Vepřek. S.: ibid, p. 492
190. Wirz, E.: Ph. D. Thesis, Univ. Zurich, 1979
191. Gimzewski, J. K., Brewer, R. J., Vepřek, S., Stuessi, H.: Proc. Fifth Inter. Symp. Plasma Chem., Edinburgh, Aug. 1981, p. 370
192. Brewer, R. J., Gimzewski, J. K., Vepřek, S., Stuessi, H.: J. Nucl. Mater., (1982) in press
193. Winters, H. F., Horne, D.: Surf. Sci. *24*, 587 (1971)
194. Lancaster, G. M., Honda, F., Fukuda, Y., Rabalais, J. W.: Chem. Phys. Lett. *59*, 356 (1978)
195. Taylor, J. A., Lancaster, G. M., Rabalais, J. W.: J. Electr. Spectr. Rel. Phenomena *13*, 435 (1978)
196. Lancaster, G. M., Rabalais, J. W.: J. Phys. Chem. *83*, 209 (1979)
197. Baldwin, D. A., Murray, P. T., Rabalais, J. W.: Chem. Phys. Lett. *77*, 403 (1981)

198. Roth, J.: In: Sputtering by Particle Bombardment, Vol. II, Ch. 3. (Topics Appl. Phys.) (Behrisch, R., ed.), Springer, Berlin—Heidelberg—New York, 1982
199. Patel, A. R., Kuruvilla, A. C.: Indian J. Pure & Appl. Phys. *19*, 803 (1981)
200. Brice, J. C.: In: Curr. Topics in Materials Sci. *4* (Kaldis, E., ed.), North-Holland, Amsterdam—New York—Oxford, 1980
201. Deryaguin, B. V., Builov, L. L., Zubkov, V., Kochergina, A. A., Fedoseev, D. V.: Sov. Phys. Crystallogr. (USA), *14*, 449 (1969)
202. Aeisenberg, S., Chabot, R.: J. Appl. Phys. *42*, 2953 (1971), J. Vac. Sci. Technol. *10*, 104 (1973)
203. Spencer, E. G., Schmidt, P. H., Joy, D. C., Sansalone, F. J.: Appl. Phys. Lett. *29*, 118 (1976)
204. a) Schmellenmeier, H.: Exp. Tech. Phys. *1*, 49 (1953); b) Z. phys. Chem. *205*, 349 (1956)
205. Freeman, J. H., Temple, W., Gard, G. A.: Nature *275*, 634 (1978)
206. Strelnitskii, V. E., Aksenov, I. I., Vakula, S. I., Padakula, V. G., Belous: Sov. Tech. Phys. Lett. *4*, 546 (1978); Sov. Phys.-Tech. Phys. *23*, 222 (1978)
207. Weissmantel, C.: J. Vac. Sci. and Technol. *18*, 179 (1981)
208. Anderson, D. A.: Phil. Mag. *35*, 17 (1977)
209. Andersson, L. P., Berg, S.: Vacuum *28*, 449 (1978); Andersson, L. P., Berg, S., Norstrom, H., Olaison, R., Towta, S.: Thin Solid Films *63*, 159 (1979); Berg, S., Andersson, L. P.: ibid., *58*, 117 (1979)
210. Holland, L., Ojha, S. M.: Thin Solid Films *38*, L17 (1976); *48*, L21 (1978); *58*, 107 (1979)
211. Gambino, R. J., Thompson, J. A.: Solid State Commun. *34*, 15 (1980)
212. Enke, K., Dimigen, H., Hubsch, H.: Appl. Phys. Lett. *36*, 291 (1980)
213. Enke, K.: Thin Solid Films *80*, 227 (1981)
214. Jones, D. I., Stewart, A. D.: J. Physique *42*, C4-1085 (1981)
215. Proc. Ninth Inter. Conf. Amorphous and Liquid Semiconductors, Grenoble, July 1981; Chakraverty, B. K.: J. de Physique *42*, 1351, C4-749 (1981); Magarino, J., Friedrich, A., Kaplan, D., Deneuville, ibid. *42*, C4-737 (1981)
216. Spear, W. E., Le Comber, P. G.: Phil. Mag. *33*, 935 (1976)
217. Wagner, J., Vepřek, S.: Plasma Chem. Plasma Proc. *2*, 95 (1982)
218. Vepřek, S., Iqbal, Z., Sarotti, F.-A.: Phil. Mag. *B45*, 137 (1982)
219. Sarotti, F.-A., Iqbal, Z., Vepřek, S.: Solid St. Commun. *42*, 465 (1982)
220. Rosner, D. E., Allendorf, H. D.: Proc. Int. Conf. Heterogeneous Kinetics at Elevated Temperatures, Univ. Pennsylvania, 1969
221. Balooch, M., Olander, D. R.: J. Chem. Phys. *63*, 4772 (1975)
222. Winters, H.: In: Plasma Chemistry III (Topics Curr. Chem. *94*) (Vepřek, S., Venugopalan, M., ed.), Springer, Berlin—Heidelberg—New York, 1980
223. Flamm, D. L., Donnelly, V. M.: Plasma Chem. Plasma Proc., *1*, 317 (1981)
224. Coburn, J. W.: ibid. *2*, 1 (1982)
225. Menzel, D.: In: Interactions on Metal Surfaces (Topics Appl. Phys. Vol. 4) (Gomer, R., ed.), Springer, Berlin—Heidelberg: 1975
226. Drinkvine, M. J., Lichtman, D.: Prog. Surface Sci. *8*, 123 (1977)
227. Madey, E., Yates, J. T.: Surface Sci. *63*, 203 (1977)
228. Chang, R. P. H., Chang, C. C., Darack, S.: J. Vac. Sci. Technol. *20*, 45 (1982)
229. Ashby, C. I. H., Rye, R. R.: J. Nuclear Materials *92*, 141 (1980); *103 & 104*, 489 (1981), published in 1982
230. Vepřek, S., Sarott, F.-A.: Plasma Chem. Plasma Proc. (in Press)

High Pressure Plasmas and Their Application to Ceramic Technology

Pierre Fauchais, Erick Boudrin, Jean Francois Coudert

Thermodynamic Laboratory — U.E.R. of Sciences
Limoges University, France

Réginald McPherson

Monash University, Australia

Table of Contents

1 Introduction

The title of this article makes reference to terms which define two domains of physics and chemistry which are apparently unconnected. However over the last three decades workers in these two disciplines have met on common research ground, mainly in the domains of aerospace and magneto-hydrodynamics. Fundamental research themes common to the two disciplines are presently in course of development, some of them proceeding to industrial trials.

High Pressure Plasmas (see Fig. 1) refer to thermal plasmas at a temperature in the range 3000–15000 K under normal pressure. Following the perfect gas law, the density ranges between $2 \times 10^{24}\ m^{-3}$ (3000 K) and $7 \times 10^{23}\ m^{-3}$ (15000 K). The equilibrium electronic energy is between 0.4 eV (3000 K) and 1.9 eV (15000 K). Using the classification proposed by Venugopalan [1] our domain of interest is then in the field of arc or high frequency, thermal plasmas.

Following Kingery [2], we define "*ceramics as the art and science of making und using solid articles which have as their essential component, and are composed in large part of inorganic, non metallic materials* ..." But in this paper we will restrict this definition to "new ceramics" such as nitrides and carbides as well as pure oxides. Some applications proposed recently in various fields, such as treatment of cements or of low grade minerals, will not be discussed.

Historically, the first applications of thermal arc plasmas were in welding and cutting (1954). Their utilisation for plasma chemistry (carbothermic reduction and chlorination of many oxide ores at first) or for plasma coating or spheroïdisation is more recent (1960–1965). The RF plasma, originally developed in 1960 as a new crystal-growing technique, revived interest in the field of plasma chemistry, owing to the absence of metal electrodes, opening the way to investigations of many chemical processes which would result in a rapid corrosion of D.C. plasma torch

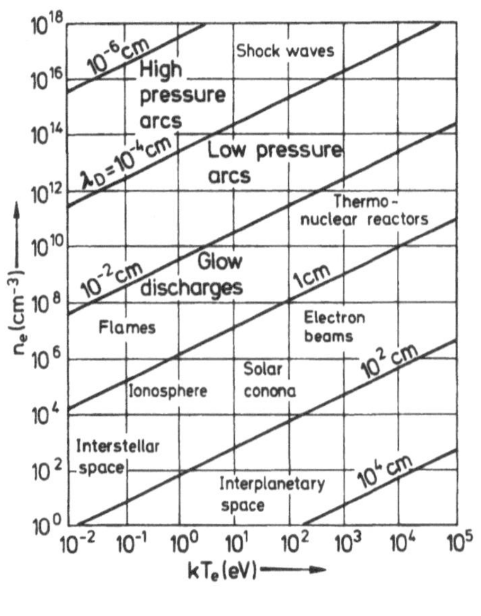

Fig. 1. Typical plasmas characterized by their electron energy and density Reprinted from [1]

component materials. These very promising technologies have been rapidly adopted in industry and have suffered, perhaps, until recently from lack of scientific interest. This situation is presently in a state of evolution and a certain number of studies, concerned with a better understanding of the basic phenomena, have been developed.

We will firstly deal with the theories and models currently used in plasma-ceramic technology: thermodynamic and transport properties, heat and mass transfer between a solid and a plasma, gas phase chemical kinetics. Then we will present a short review of the plasma measurement techniques (mainly temperature and velocity).

In the third part, we will briefly describe the main experimental devices used to produce thermal plasma.

The fourth part will give a review of physical transformations undergone by ceramics under plasma conditions (fusion, evaporation, condensation). Concerning plasma spraying of ceramics, particular attention will be devoted to the comparison of models with real systems and spheroïdisation will be treated simultaneously with spraying.

In the sixth part, concerning synthesis (or to a lesser extent reduction) of ceramics in plasmas we shall include some of the experimental knowledge acquired up to the 1970's which has been reviewed by Stokes [3]. Although certain experiments reported in his work were very promising, a certain disinterest in this field has been apparent until recently. However this situation has changed during the last years because of a significant progress which has been achieved in understanding of fundamental processes as well as in applications.

2 Theories and Models in the Field of Ceramic Plasma Technology

2.1 Equilibrium Theories

The plasma systems most often used in the domain of ceramics make use of physico-chemical reactions in a plasma stream. Under ideal conditions of thermodynamic equilibrium it is possible to determine the products resulting from a given treatment. Furthermore knowledge of the chemical composition of a plasma at a given temperature and pressure is needed to perform the calculation of the classical thermodynamic properties such as the specific heat, and the enthalpy as well as the transport properties: electrical and thermal conductivities and viscosity. The problem is to know to what extent calculations assuming complete thermodynamic equilibrium are applicable to real systems. In the case of a plasma stream, thermodynamical equilibrium can be applied if the chemical reactions proceed more rapidly than the physical phenomena such as the relaxation of energy levels or the maxwellisation of velocity. The detailed study of this type of phenomenon is outside the scope of this article, but it is possible by means of simplified considerations to find the orders of magnitude of reaction times corresponding approximately to equilibrium conditions. Along the axis of a D.C. plasma jet the temperature varies by 1% in 2.5×10^{-5} m, so that with a flow velocity of 250 m/s this variation of temperature of 1% corresponds to 10^{-7} s. In order to attain chemical equilibrium in that time interval, the reaction times must be smaller or equal to 10^{-7} s. Some recent kinetic calculations (discussed later) show that this reaction time exists at temperatures above 5000 K, in most cases of plasma chemistry one can therefore apply equilibrium results as a first approach in the high temperature jet, but probably not in the low temperature regions. Moreover, if the velocity of the plasma jet is sufficiently high (more than 800 m/s) the heating rate is great enough (10^9 K/s) to modify the equilibrium conditions in the hot core of the plasma as we have showed for nitric oxide synthesis in D.C. plasma jets.

However, these considerations do not apply in the case of heterogenous reactions between the plasma and solid particles. A solid particle is actually at a temperature very much lower than the plasma and the rate of mass transfer between the plasma and the particle is infinitely slow compared with the reaction rate in the gas phase. Nevertheless thermodynamic calculations are often applied at the level of the condensed phase giving results which may be regarded as an "upper limit" for the extent of reaction.

2.1.1 Determination of the Equilibrium Chemical Composition

The concentrations in an elementary volume of plasma are generally determined at fixed temperature and pressure. The method consists in finding the chemical composition which gives the minimum of the Gibbs free energy [4] under the constraints of the laws of conservation of matter and electrical neutrality.

$$\text{Formally:} \quad \begin{cases} G = G(T, p, n_i) \text{ minimum} \\ \sum\limits_{l} \alpha_{li} n_i - B_l^0 = 0 \end{cases}$$

where T is the temperature, p the pressure, n_i the number of moles of the ith specie, α_{li} the number of gram atoms of element 1 in the specie i and B_l^0 the number of gram atoms of the element l present in the closed system considered.

Under high temperature conditions, the major part of the polyatomic species are dissociated and the number of coexisting species generally becomes large. It is hardly possible to solve the system of non linear equations in closed form, so one must use a numerical method on computer. A review of these numerical methods can be found in the monograph of Storey and Van Zeggeren [5] as well as FORTRAN IV source programs by Bourdin [6].

The Gibbs free energies are calculated using standard thermodynamic tables which are easily usable by machine since they give the data in the form of polynomial coefficients. The data are sometimes limited to 6000 K and it is therefore necessary to make extrapolations or to carry out calculations of partition functions from spectroscopic data [7]. In the latter case, which is certainly more reliable, one can determine standard thermodynamic functions with the aid of the classical formulae of statistical thermodynamics. The results may then be fitted to polynomials so that they match the tabulated data [6]. Furthermore the calculation of partition functions is necessary for spectroscopic diagnostics and for the calculations of reaction rate parameters.

2.1.2 Determination of Thermodynamic Functions

When, at equilibrium, the chemical composition of a mixture is known, the thermodynamic properties such as specific heat or enthalpy can be calculated. We shall look firstly at specific heat at constant pressure, C_p, defined by:

$$C_p = \left(\frac{\partial \hat{h}}{\partial T}\right)_p$$

where $\hat{h}$ is the specific mass enthalpy of the mixture composed of N chemical species i

$$\hat{h} = \sum_{i=1}^{N} x_i H_i \Big/ \sum_{i=1}^{N} x_i m_i$$

x_i is the mole fraction, H_i the enthalpy and m_i the molar mass of species i.

Calculation of the partial derivative gives [8]:

$$C_p = \frac{1}{M}\left(\sum_{i=1}^{N} x_i C_{pi} + \sum_{i=1}^{N} x_i (H_i - m_i \hat{h}) \left(\frac{\partial \operatorname{Log} x_i}{\partial T}\right)_p\right)$$

where C_{pi} is the specific heat at constant pressure of the species i, and M the total mass of the mixture. This expression for C_p is simply the sum of two terms called respectively the frozen $\bar{C}_{pf}$ and the reactional specific heat C_{pr}:

$$C_p = \bar{C}_{pf} + C_{pr}$$

The frozen term represents the sum of the contributions of each of the species, whereas the reaction term is associated with chemical reactions at the temperature considered. The influence of this second term is predominant as we shall show from the results obtained for nitrogen and the carbon-hydrogen-oxygen system. The expression for $C_v = \left(\frac{\partial \hat{h}}{\partial T}\right)_v$ is similarly $C_v = \bar{C}_{vf} + C_{vr}$, the reaction term being again predominant. When the data for the system are tabulated, calculation of the frozen term is not a problem; on the other hand, calculation of the reaction term is more difficult, since it is necessary to use the laws of conservation in addition to the equilibrium conditions which make use of the law of mass action in the case of perfect gases. It is then much more simple to perform calculations of a numerical derivative of the enthalpy $\hat{h}$. The calculation of C_p, C_v and of the sound velocity $v_s = \left(\frac{RT\Gamma}{m}\right)^{1/2}$ may be calculated at the same time [4] as C_p by calculation of the isentropic coefficient $\Gamma = \left(\frac{\partial H}{\partial V}\right)_s$ computed with $\left(\frac{\partial \operatorname{Log} x_i}{\partial T}\right)_p$. The sound velocity is used in flow calculations.

The enthalpy and entropy of the mixture are obtained by integration of the specific heat at constant pressure if these quantities are tabulated:

$$\hat{h} - \hat{h}_{T_0} = \int_{T_0}^{T} C_p(T)\, dT \qquad S = \int_{0}^{T} \frac{C_p(T)}{T}\, dT$$

In case of lack of these tabulated data it is then necessary to calculate the enthalpy and the entropy directly using either the first derivative of the partition function or the values of the enthalpy and entropy of the different species which are more often tabulated.

2.1.3 Examples

2.1.3.1 Nitrogen Plasma

Many calculations and measurements have been performed on nitrogen (for example Drellishak)[9] which is the gas mostly used in plasma torches; it is therefore of interest to deal with this gas as the first example. Figure 2 gives the equilibrium composition of nitrogen at one atmosphere, as a function of temperature. This diagram shows the effect of temperature on dissociation and ionisation. Figure 3 shows the

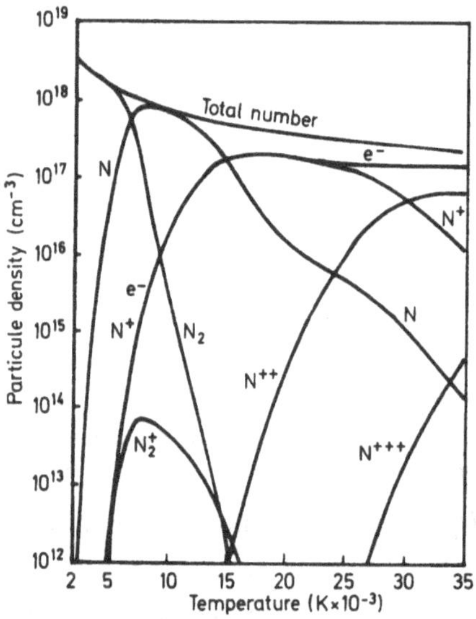

Fig. 2. Equilibrium composition versus temperature of a nitrogen plasma at 1 atm. Reprinted from [9]

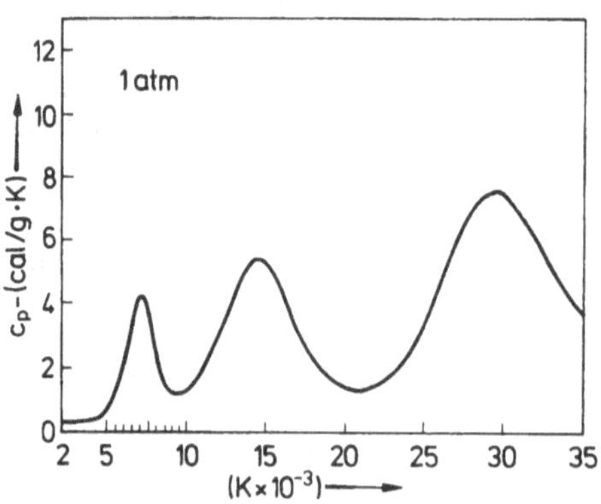

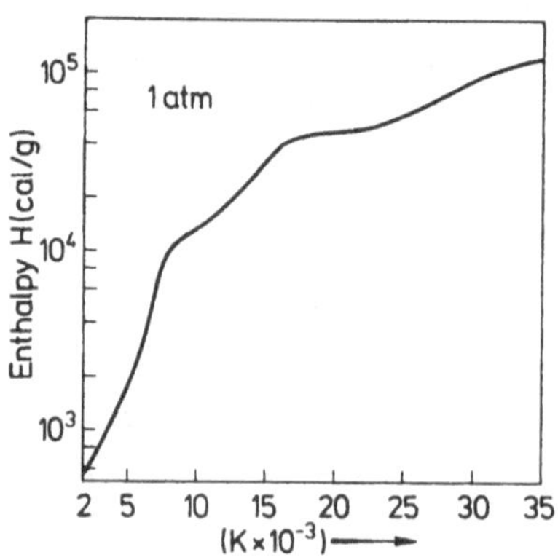

Fig. 3. Specific heat C_p versus temperature of a nitrogen plasma at 1 atm. Reprinted from [9]

Fig. 4. Enthalpy change versus temperature of a nitrogen plasma at 1 atm. Reprinted from [9]

function $C_p(T)$; the two maxima between 2000 and 20000 K are the reactional components C_{pr} of the specific heat. It is clear that these peaks are due to the dissociation of N_2 and ionisation of N respectively. Figure 4 represents the enthalpy of the plasma versus temperature, obtained by integration of the preceeding curve; it is also very clear that the steep variations of this enthalpy are essentially due to the heats of reaction (dissociation, then ionisation). It is very important to emphasize that these heats of reaction lead to large variations of the thermodynamic properties such as specific heat or enthalpy or entropy making impossible to extrapolate the data, usually well known only up to 2000–2500 K. We shall also find similar wide variations in thermal conductivity arising from the same effects.

2.1.3.2 Nitrogen-Aluminium Plasma, Nitrogen-Silicon Plasma

Calculations [6] of these systems have been motivated by a study concerning the preparation of nitrides from these elements in a nitrogen plasma. The processes examined for this synthesis were reaction in the vapour phase followed by condensation of the products, the original idea being to accelerate, by high temperature heating followed by a fast quenching, the reactions of formation of AlN or Si_3N_4. It may be seen from the equilibrium diagrams (Fig. 5) that due to dissociative evaporation, the maximum concentration of AlN(g) or SiN(g) is low ($\sim 10^{14}$ cm^{-3}). However, as we shall see in the last part, it is possible to obtain a high conversion into AlN(s) from elements. A possible explanation is that under the particular experimental conditions the recombination reaction Al(g) + N → AlN(s) proceeds faster than dissociative evaporation AlN(g) → Al(g) + N. The evidence from this example shows that equilibrium calculations may be misleading because the quenching kinetics is the determining step. This statement is rather general for reactions of the

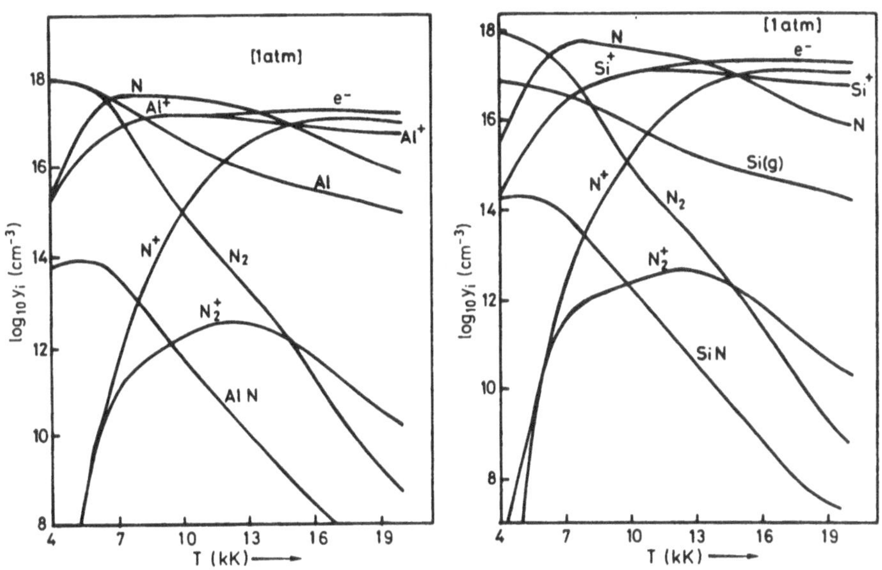

Fig. 5. Equilibrium composition versus temperature of a N_2—Al mixture and of a N_2—Si mixture at 1 atm. Reprinted from [6]

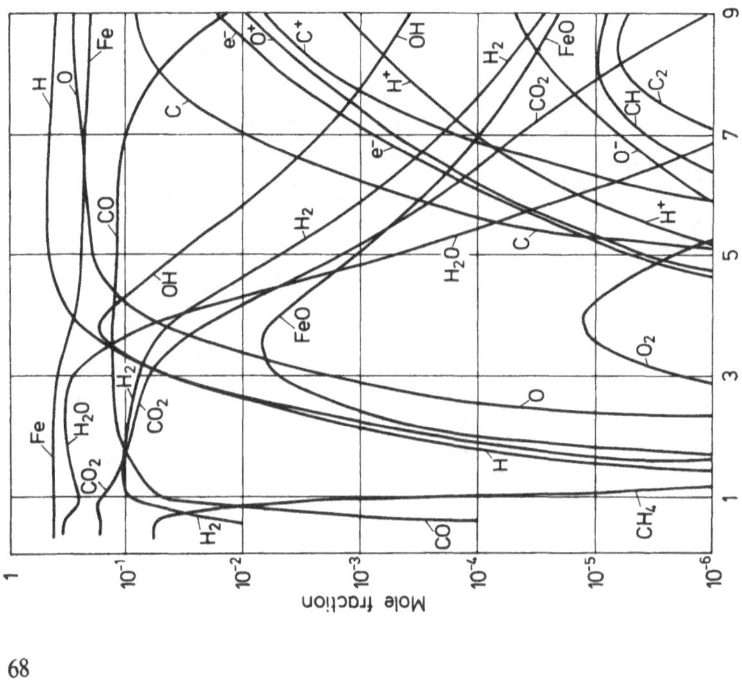

Fig. 6. Equilibrium composition of a mixture of (C, H, O, Fe) (1 : 4 : 3 : 2) at 1 atm. Reprinted from [8)]

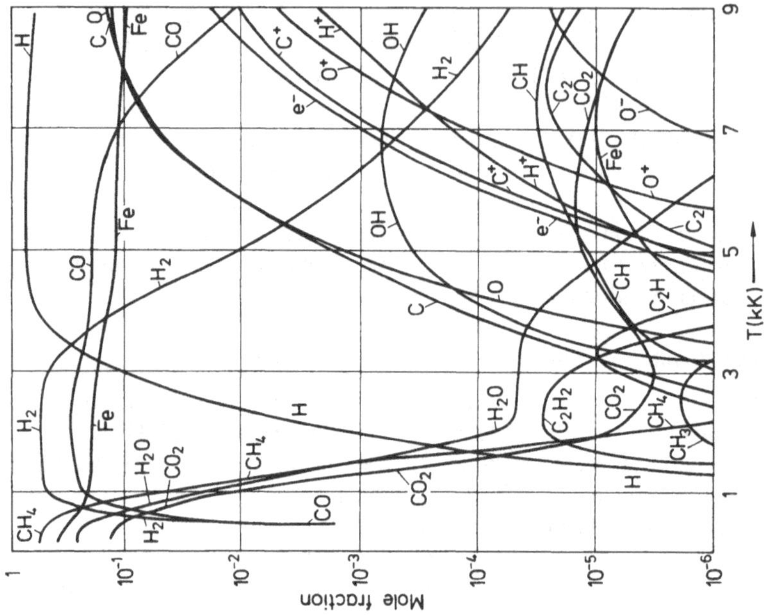

Fig. 7. Equilibrium composition of a mixture of (C, H, O, Fe) (3, 12, 3, 2) at 1 atm. Reprinted from [8)]

type gas — solid for which calculation of the equilibrium concentrations in the gas phase may be of secondary importance in providing an estimate of the extent of reaction because of the influence of processes occuring during condensation on the overall kinetics.

2.1.3.3 Plasmas C, H, O, Fe

Figure 6 gives the mole fractions at equilibrium for the system 1 C, 4 H, 3 O, 2 Fe. This mixture ($CH_4 + Fe_2O_3$) has been studied [8] to provide data on the equilibrium vapour phase composition in a plasma furnace used for the reduction of iron oxide. A priori this diagram shows that the reduction of iron oxide to iron attains 100% in the vapour phase. As we have shown previously for the nitrides the problem of condensation remains unresolved, however, experimental results indicate that the cooling rate has little effect on the overall yield in that particular case. Figure 7 shows the reducing powers of hydrogen and carbon.

Figure 8 shows the different components of the specific heat at constant pressure for the system 1 C, 4 H, 3 O, 2 Fe. The very reactive nature of the mixture is indicated by the form of the C_p versus temperature curve.

The enthalpy and entropy changes of the same system are given in Fig. 9; the curves are analogous to those presented for nitrogen and have the same orders of magnitude. Except for the reducing nature of this mixture, it may be seen that its behaviour is similar to that of nitrogen. We shall see for the particular case of hydrogen however, that the thermal conductivity, a basic parameter in heat transfer, completely changes a situation which would be based purely on considerations of specific enthalpy.

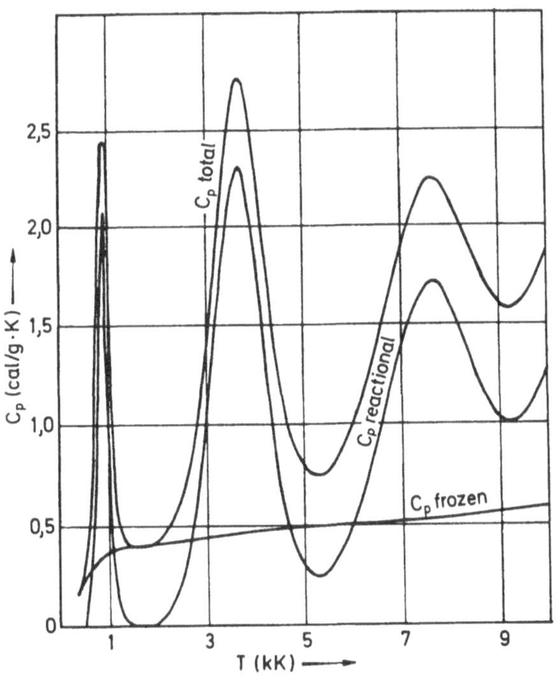

Fig. 8. Constant pressure specific heat C_p change versus temperature of the mixture (C, H, O, Fe) (1:4:3:2) at 1 atm. Reprinted from [8]

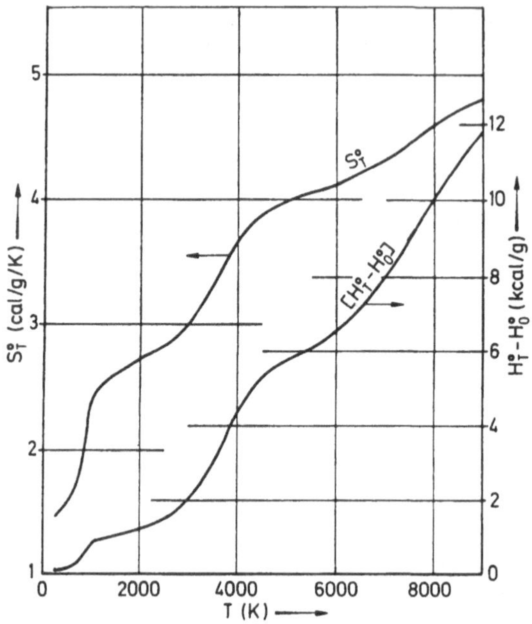

Fig. 9. Enthalpy change versus temperature of the mixture (C, H, O, Fe) (1:4:3:2) at 1 atm. Reprinted from [8)]

2.1.3.4 Study of Mixtures of SiO_2 and C

This study provides an interesting example of a solid gas system. Calculations have been made by Coldwell [10)] on the system SiO_2—C from the point of view of reduction of SiO_2 to silicon. Figure 10 which illustrates the results of this study, shows the existence of a threshold temperature in the range 2400–2500 K. On one side of this temperature the only species are SiO_2, SiO and SiC which are dissociated to a great extent above 2500 K. The author indicates that since the calculations are done only for the vapor phase, the threshold temperature is higher and that it will be lower than 2400 K if acetylene rather than carbon is used as reducing agent. The reduction (from the point of view of equilibrium) proceeds in part in the condensed phase, this point being favorable in plasma systems in which it may be difficult to assume complete vaporisation of SiO_2. It is worth noting from Fig. 10 that there is always a low concentration of SiO above the threshold temperature, but thermodynamics cannot describe this effect during the cooling phase.

2.1.3.5 Conclusion

The analysis of the examples concerning thermodynamic equilibrium seems to be characteristic of the plasma-ceramics domain. We have tried to put forward the advantages and disadvantqges of the use of chemical thermodynamics as applied to this area and we shall see in the following sections that many studies are based on the same point of view.

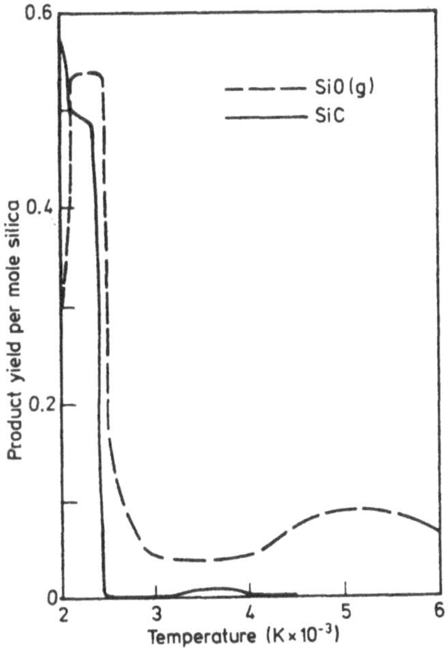

Fig. 10. Production yield of SiO and SiC at equilibrium versus temperature for the system (SiO_2, 2 C). — Reprinted from [10]

2.2 Transport Properties

When a plasma is considered as a source of energy to heat solid materials and to produce physical and chemical changes, the level of heat transfer between the plasma and introduced material assumes considerable importance. As we shall see in the section devoted to transport between plasma and solid phases, the parameters which have the major influence are the *viscosity* and *thermal conductivity* of the plasma. When systems are used which employ electric arcs, the *electrical conductivity* of the gas is obviously a property which cannot be ignored. Furthermore, any study of a plasma process based on reactions in the vapour phase implies that the diffusion coefficients of the reacting species are known since these control the mixing of the plasma and vapour. The parameters viscosity. diffusion coefficient, electrical and thermal conductivity all characterise transport of energy or material within the plasma fluid and are generally treated under the heading of transport properties. This section is concerned with theoretical means available to determine those transport properties which are difficult to measure experimentally [11]. Some examples from currently known systems allow the theory to be illustrated and will be used in the following sections.

2.2.1 Calculation of Transport Properties

For plasma in local thermodynamic equilibrium (L.T.E.), calculation of the transport properties begins with the kinetic theory of gases. This theory is based on statistical concepts, in particular, on the *distribution function* of the gas particles [12]. For a particle of the i-th type, the distribution function $f_i(\vec{r}, \vec{v}_i, t)$ represents the number of particles of type i present, at the instant t, in a elementary colume $d\vec{r}$, $d\vec{v}$ of phase space $\vec{r}$, $\vec{v}$. At equilibrium, the distribution function is the same for all particles; it is the well known Maxwell function:

$$f_i^o(\vec{r}_i, \hat{v}_i, t) = n_i \left(\frac{m_i}{2\pi kT}\right)^{3/2} e^{\left(-\frac{m_i v_i^2}{2kT}\right)}$$

where T is the equilibrium thermodynamic temperature. The Maxwell distribution is independent of time and of direction of $\vec{v}_i$. In the general case, the function $f_i(\vec{r}, \vec{v}, t)$ is a solution of the integro-differential Boltzmann equation which establishes a balance for particles in an elementary volume of phase space:

$$\frac{\partial f_i}{\partial t} + \vec{v}_i \, \vec{\text{grad}}_r \, f_i + \frac{\vec{F}_i}{m_i} \, \vec{\text{grad}}_{v_i} \, f_i = \left(\frac{\partial f_i}{\partial t}\right)_{\text{coll}}$$

The first term of the left hand side (L.H.S.) takes into account the explicit variation of f_i with time, the second term the variation of f_i due to displacement of particles and the third term displacement of particles under the action of external forces. The R.H.S. term takes account of the collisions.

If the function $f_i(\vec{r}, \vec{v}_i, t)$ is known it is possible to calculate, from the statistical point of view, the mean values of all quantities of the type $\alpha(\vec{r}, \vec{v}_i, t)$:

$$\alpha(r, t) = \frac{1}{n_i(r, t)} \int \alpha(\vec{r}_i, \hat{v}_i, t) \, f_i(\vec{r}_i \vec{v}_i t) \, dv_i$$

In particular the *heat flux vector*

$$\vec{Q}(\vec{r}, t) = \sum_i \frac{1}{2} \, n_i m_i \, \overline{|\vec{v}_i - \vec{v}_0|^2 (\vec{v}_i - \vec{v}_0)}$$

and the *pressure tensor*

$$\bar{\bar{P}} = \sum_i n_i m_i \, \overline{(\vec{v}_i - \vec{v}_0)(\vec{v}_i - \vec{v}_0)}$$

can be determined, where $\vec{v}_0$ is the mean mass velocity. The right hand side (R.H.S.) of the Boltzmann equation takes account of elastic collisions only, thus the calculations will only be suitable for unionised and unexcited monatomic gases. The theory must therefore be modified for thermal plasmas.

To solve the Boltzmann equation [13,14]. Enskog's method of successive approximations is generally used [15]. This consists of writing firstly,

$$f_i = f_i^0(1 + \varphi_i)$$

where φ_i is a perturbation function which allows passing from zero order (Maxwell-Boltzmann) to first order. φ_i is found [16] by using a Sonine polynominal series of order k. The complcxity of the calculation increases considerably with increasing values of k, it is thus necessary to find the minimum value of k which gives the best convergence of the series.

Athye [17] and Devoto [18,19,20,21] have shown that, for thermal conductivity, it may be necessary to carry out calculations up to $k = 4$ in the cases of hydrogen, argon, helium and krypton. One often obtains simplified expressions for the Boltzmann equation by treating electrons and heavy particles separately since electron-heavy particle collisions produce little change in the perturbation function of the heavy particles. Devoto [13] has shown that, for the thermal conductivity, this simplification introduces an error of less than 0.1% for argon and less than 0.9% for nitrogen if the terms for heavy particles are calculated using $k = 2$ and that for electrons, $k = 7$.

In addition to the problems arising from expansion of the polynomials, collision integrals make their appearance within the framework of the theory of Chapman [16]. These are calculated from the interaction potentials between the different particles. The collision integrals are generally only known for the "simple gases" (nitrogen, oxygen, argon, hydrogen); unfortunately the data are unknown in other cases, for interactions of the type C—H for example (methane-hydrogen plasmas are often used in plasma chemistry). The diffusion coefficients, involved in the expressions for transport coefficients are strongly connected to the collision integrals and the effect of uncertainty in the interaction potential may be seen. For example, Gorse [2] has shown that the choice of potentials for $N—N^+$ can introduce uncertainty of 60% and 75% in the thermal conductivity and viscosity of nitrogen respectively between 10000 and 20000 K.

2.2.1.1 *Viscosity*

The viscosity coefficient establishes the proportionality between the friction force in the direction of flow and the velocity gradient in the orthogonal direction. Calculation of this coefficient is made with the pressure tensor in the first approximation of Chapman-Enskog [13].

2.2.1.2 *Thermal Conductivity*

The distribution function and the heat flux vector allows the *translational thermal conductivity* to be determined. The latter can be written as the sum of two terms, one due to the electrons $\varkappa_{tr}^{e}$ and the other due to heavy particles $\varkappa_{tr}^{h}$. The two terms are calculated in the 3rd and 2nd approximation of Chapman-Enskog, respectively. We have seen in the section that, at high temperatures, the gas is dissociated and ionised. The dissociation and ionisation phenomena make a large contribution to the energy transport. Considering a diatomic gas, it is dissociated and ionised in the high temperature regions of the plasma and recombination occurs when it diffuses to the cooler regions and vice versa. This results in transport of chemical energy which is taken into account as *reaction conductivity*. This term is determined by means of the theory of Butler and Brokaw [23] exclusively using the hypothesis that $k = 1$ for reasons of symmetry.

Since each species at high temperatures has an internal energy component (vibration, rotation, electronic excitation) energy is transfered in the course of an inelastic collision when the particle becomes unexcited. This effect is taken into account by the *internal thermal conductivity* which is determined from the theory of

Euken[24], also using the first approximation. The total thermal conductivity is considered to be the sum of these three contributions $\varkappa = \varkappa_{tr} + \varkappa_{re} + \varkappa_{int}$. The respective contribution of these three terms will be considered for the example of nitrogen which we have studied in detail from the point of view of equilibrium.

2.2.1.3 Electrical Conductivity

In the presence of an electric field in the arc, or under the action of other phenomena such as gradients of temperature, pressure or concentration, transport within the plasma occurs by electric charges which constitute an electric current. The current density can be written:

$$j = \sum_i z_i n_i \vec{\bar{V}}_i$$

where z_i, n_i and $\vec{V}_i$ are respectively the charge, density and diffusion velocity of the k_i ionised species. The mean diffusion velocity $\bar{V}_i$ can be calculated with the aid of the distribution function f_i. Thus the electrical conductivity σ (such that $\vec{j} = \sigma\vec{E}$) is expressed as a function of the diffusion coefficient which can be calculated for different orders of k (generally k = 3)[13].

2.2.2 Examples

2.2.2.1 Viscosity

To illustrate the preceeding theory we have represented in Fig. 11 the viscosity of Argon, Nitrogen, Hydrogen, Helium[22,25], as a function of temperature. The viscosity of plasmas at 10000 K lies in the range $1 \cdot 10^{-3}$ (g/cm · s) — $4 \cdot 10^{-3}$ (g/cm · s). These values are approximately ten times higher than those of the same gases at room temperature and the consequences of this fairly high viscosity will be seen in

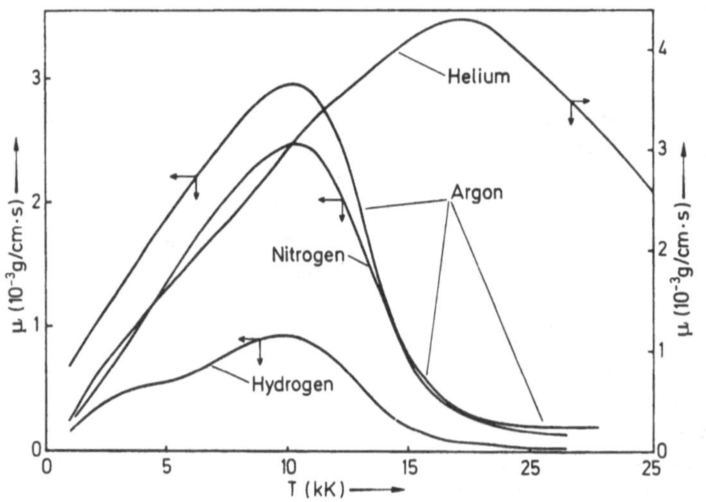

Fig. 11. Viscosity of hydrogen, nitrogen, argon and helium plasma at one atm. From the results of[22] and[25]

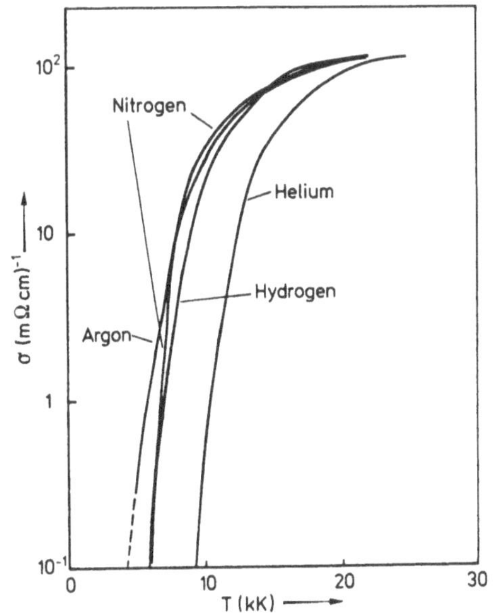

Fig. 12. Electrical conductivity of hydrogen, nitrogen, argon and helium plasma at one atm. [22)]

the following paragraphs. The viscosity increases with temperature up to a maximum at about 10000 K for Argon, Nitrogen, Hydrogen, and at about 17000 K for Helium, and then decreases drastically to very low values. This drop in viscosity is due to the ionisation of the gas; the forces between particles are coulomb forces of relatively long range, the mobility is decreased and this reduces the transport of momentum [25)]. The influence of ionization may be emphasized by comparing the ionization energy of N, H, and Ar, which are of the same order ($\sim$ 14 eV) with that of Helium (24,6 eV).

2.2.2.2 Electrical Conductivity

The influence of ionization on the electrical conductivity is quite obvious; the plasma is an electrical conductor when the gas is sufficiently ionized. Ar, N_2 and H_2 wich have a comparable ionization energy have consequently about the same conductivity (σ) as a function of temperature; at 10000 K σ is 30 Ω^{-1} cm^{-1}. At the same temperature He has a very low σ of 5 Ω^{-1} cm^{-1} due to its high ionization energy. Figure 12 shows the curves σ(T) for He, Ar, N_2 and H_2. The conductivity increases sharply with temperature and reaches a limiting value at 10^2 Ω^{-1} cm^{-1} until second ionization occurs. Finally it must be noted [27)] that the electrical conductivities of the mixtures Ar—N_2, Ar—H_2, N_2—H_2 do not depend on the ratio of the mixture.

2.2.2.3 Thermal Conductivity

As we have seen in the preceding paragraphs the thermal conductivity is separated into three parts corresponding to translational, reactional and internal conductivity ($\varkappa_t$, $\varkappa_{re}$, $\varkappa_{int}$). Figure 13 shows these three contributions for nitrogen; translational and internal conductivities of neutrals give a quite small contribution to the total

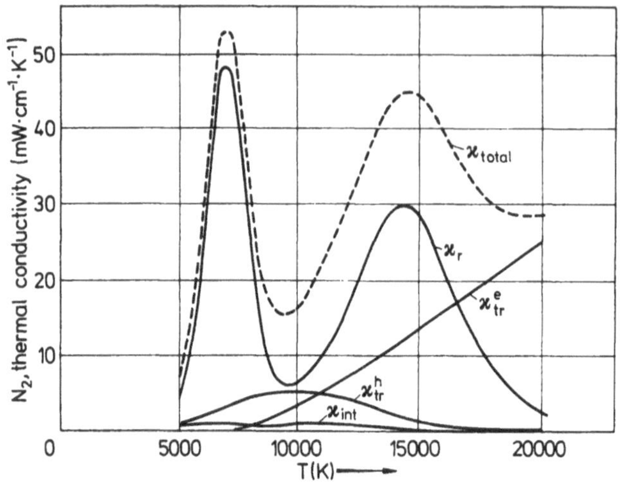

Fig. 13. Translational (electrons and heavy species), reactional, internal and total thermal conductivity of a nitrogen plasma at one atm. Reprinted from [22]

conductivity. The main contribution is clearly the reactional one composed of two maxima due to the dissociation and ionization energies respectively, (the same kind of reaction contribution has been discussed for specific heat see Sect. 2.1.3.1). At high temperatures the translational conductivity of electrons becomes significant. It is interesting to compare Fig. 13 to Fig. 14 [22] which shows the thermal conductivity of argon as a function of temperature. At low temperature there is obviously no dissociation and the translational contribution of neutral species is the most significant: the reactional part begins to be important at 10000 K together with the contribution of the electron because Ar is significantly ionized. At very high

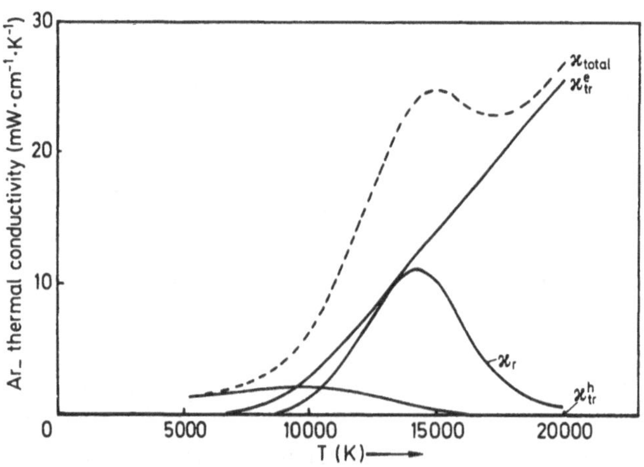

Fig. 14. Translational (electrons and heavy species), reactional, internal and total thermal conductivity of an argon plasma at one atm

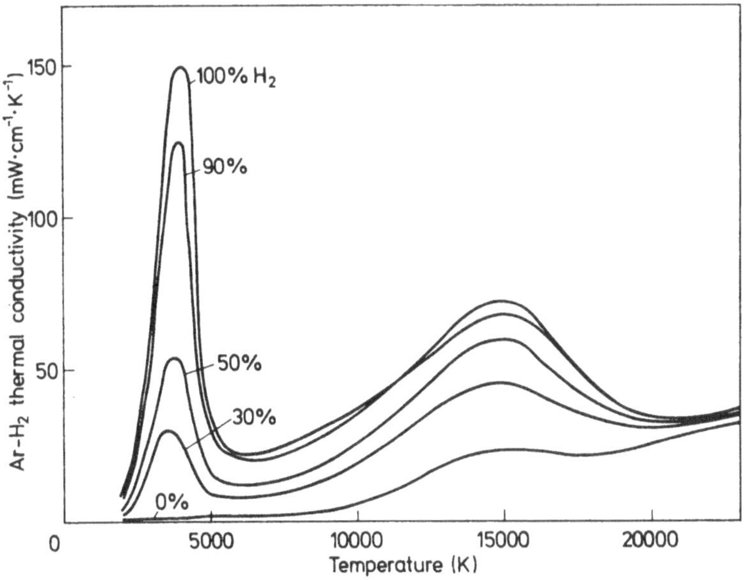

Fig. 15. Thermal conductivity of a mixture of hydrogen and argon at one atm. Reprinted from [22]

temperatures when the first ionization is completed, the translational conductivity of electrons is the main contribution until second ionization occurs.

Figure 15 shows the thermal conductivity [22] of a mixture of hydrogen and argon. Hydrogen has a very low dissociation energy (4.48 eV) compared to nitrogen (9.76 eV) so that the reactional conductivity (due to dissociation) occurs at quite a low temperature and gives a very high thermal conductivity. The reactional (ionization) contribution at high temperature is almost comparable to that of nitrogen (ionization energies are of the same order).

Figure 16 shows the influence of the uncertainties on the interaction potentials for the thermal conductivity of hydrogen. It should be pointed out that uncertainties of a factor 2 can occur for the translational conductivity due to uncertainties of the interaction potentials $U(H_2—H_2)$. The same kind of error results for the other transport properties. The great importance of thermal conductivity in the utilization of plasma for material treatment will be extensively studied in the next section.

2.3 Heterogeneous Models

Before dealing with the problems of heterogeneous heat and mass transfer (plasma-wall or plasma-particle) we shall recall some of the basic relations used in that field. The reader familiar with fluid mechanics formulations can proceed directly to Section 2.3.2.

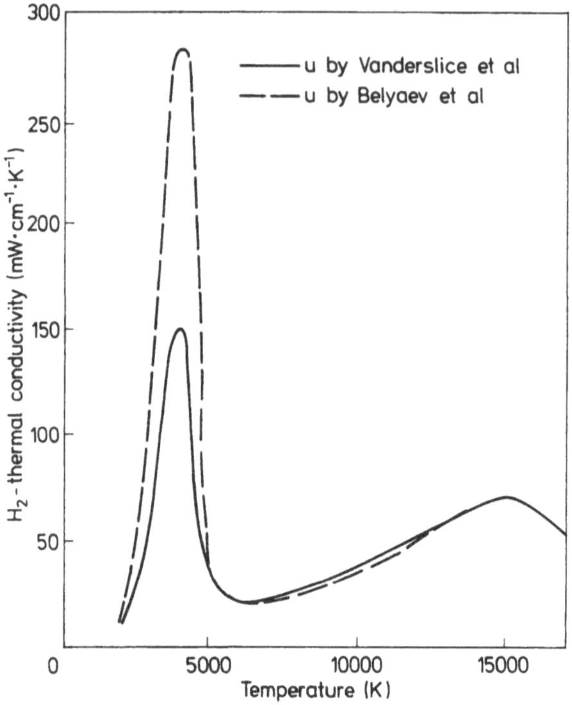

Fig. 16. Influence of the choice of the H—H interaction potential on the calculated thermal conductivity of an hydrogen plasma at one atm. Reprinted from [22]

2.3.1 General Considerations

2.3.1.1 Balance Equations for an Homogeneous Reactive Flow

The following relations are those governing reactive flows controlled by diffusion, heat transfer and momentum transfer phenomena [26–34].

a) *Mass conservation*

Let ϱ_j be the mass density of species j moving with a velocity $\vec{v}_j$ ($\vec{v}_j$ is the sum of diffusion velocity $\vec{V}_j$ and center of mass velocity $\vec{u}$).

The conservation equation is:

$$\frac{\partial \varrho_j}{\partial t} + \nabla \cdot (\varrho \vec{v}_j) = \omega_j \tag{1}$$

where ω_j is the mass production rate by chemical reactions. The continuity equation is obtained by summing (1) over all the chemical species.

$$\frac{\partial \varrho}{\partial t} + \nabla \cdot (\varrho \vec{u}) = 0 \tag{2}$$

The diffusion flux of species j is defined by:

$$\vec{J}_{Dj} = \varrho_j \vec{V}_j \tag{3}$$

The convective flux of species j is defined by:

$$\vec{J}_{cj} = \varrho_j \vec{u} \tag{4}$$

b) *Momentum conservation*

We first define the time derivative operator as:

$$\frac{d}{dt} = \frac{\partial}{\partial t} + \vec{u} \cdot \nabla \tag{5}$$

For viscous flow, momentum conservation can be written as:

$$\varrho \frac{d\vec{u}}{dt} + \nabla p = -\nabla \cdot \vec{\Pi} \tag{6}$$

where $\vec{\Pi}$ is the viscous stresses tensor defined by the following expression:

$$\vec{\vec{\Pi}}_{ik} = \mu \left(\frac{\partial u_i}{\partial X_k} + \frac{\partial u_k}{\partial X_i} \right) - \eta (\nabla \cdot u)\, \delta_{ik} \tag{7}$$

μ is the viscosity coefficient; η is the second viscosity coefficient used for compressible flow.

c) *Energy conservation*

Under the ethalpy formulation, the balanced equation becomes:

$$\varrho \frac{d\hat{h}}{dt} = \frac{d\varrho}{dp} + \varphi - \nabla \cdot \vec{q} \tag{8}$$

$\hat{h}$ is the unit mass enthalpy of the fluid, φ is the viscous dissipation term defined as:

$$\varphi = -\Pi : \nabla \vec{u} \tag{9}$$

$\vec{q}$ is the heat flow given by the expression [35]:

$$\vec{q} = \varkappa \nabla T + \vec{q}_R + \Sigma_j \varrho_j \hat{h}_j \vec{V}_j + \frac{RT}{\varrho} \sum_j \sum_i \frac{\varrho_i D_{T,j}}{M_j D_{ji}} (\vec{V}_j - \vec{V}_i) \tag{10}$$

where the phenomena considered are heat conduction, radiative flux, enthalpy diffusion and heat transfer by the Duffour effect respectively

$D_{T,j}$ is the thermal diffusion:. $D_{T,j} = \left(\frac{\partial \operatorname{Log} n_j}{\partial \operatorname{Log} T} \right)_P$

D_{ji} is the linear diffusion: $D_{ji} = \frac{X_j X_i}{M_{ji} v_{ji}} P$

where n_j is the mole number of the specie j, X_j, X_i are molar fractions, M_{ji} is the reduced mass of species j and i and ν_{ji} is the collision frequency between species j and i.

2.3.1.2 Transfer Coefficients

In most cases, the solutions of the problems set up for heat, mass and momentum transfer evaluations in convective flows cannot be obtained by analytical means. Generally, approximations are used based on empirical relations or coefficients.

Considering a unit surface, the quantity of heat passing through this surface in unit time is given by the expression:

$$q_0 = (\vec{p} \cdot \vec{n})_0 \tag{11}$$

where $\vec{n}$ is the normal to the surface.

Subscript o means that this quantity is evaluated at the interface.

An analogous relation is given for mass transfer:

$$j_0 = (j_{\varrho} \cdot \vec{n})_0 \tag{12}$$

or for momentum transfer:

$$\vec{\tau}_0 = (\vec{\tau} \cdot \vec{n})_0 \tag{13}$$

$\vec{q}$, j_D, $\vec{\tau}$ are respectively heat flow vector, mass flux vector and momentum flux tensor

q_0, j_0, $\vec{\tau}$ lead to three local coefficients [36]:

Heat transfer coefficient:

$$h_T = \frac{q_0}{T - T_0} \tag{14}$$

Mass transfer coefficient:

$$h_\omega = \frac{j_0}{\varrho - \varrho_0} \tag{15}$$

Friction coefficient:

$$\vec{f}_0 = \frac{\vec{\tau}_0}{\frac{1}{2}\varrho u^2} \tag{16}$$

$(T - T_0)$ and $(\varrho - \varrho_0)$ are characteristic temperature and mass concentration differences in the neighbourhood of the interface, respectively and $\frac{1}{2}\varrho u^2$ is fluid dynamic pressure.

2.3.1.3 Similitude Notions

The foundation of similitude theory is that there is no law of nature depending on the choice of units used for describing the corresponding phenomenon. Consequently, these laws can be expressed by relations between dimensionless quantities termed "similitude parameters" (see Table 1).

a) *Characteristic dimensionless group*

The major problem is determination of heat and mass transfer coefficients. Similitude theory allows the definition of two dimensionless coefficients including h_T and h_ω. These are the so called Nusselt and Stanton numbers given by:
For heat transfer:

$$Nu = \frac{h_T d}{\varkappa} \tag{17}$$

$$St = \frac{h_T}{u \varrho c_p} \tag{18}$$

For mass transfer:

$$Nu_\omega = \frac{h_\omega d}{D} \tag{19}$$

$$St_\omega = \frac{h_\omega}{u} \tag{20}$$

Note that Nusselt number for mass transfer is usually known as Sherwood number. That is:

$$Nu_\omega = Sh \tag{21}$$

Table 1. Dimensionless quantities

— *Coordinates and time*:

$x_i^* = x_i/D$; $t^* = t|\vec{u}|/D$

— *Differential operators*:

$\nabla^* = D.\nabla$
$\nabla^{*2} = D^2\,\nabla.\nabla$

$$\frac{d}{dt} = \frac{D}{|\vec{u}|}\left(\frac{\partial}{\partial t} + \vec{u}\nabla\right)$$

— *Linked variables*:

$$\vec{u}^* = \frac{\vec{u}}{|u|};\quad \varrho^* = \frac{\varrho}{\varrho_0};\quad T^* = \frac{T - T_0}{\Delta T};\quad Y_i^* = \frac{Y_i - Y_{i0}}{\Delta Y_i}$$

$$\varphi^* = \frac{1}{u}\frac{D^2}{|u|^2}\varphi;\quad P^* = (\varphi^* - \varphi_0^*) + \frac{P - P_0}{\varrho|u|^2}$$

D is a characteristic dimension of the problem, ϱ_0, T_0, Y_{i0}, P_0 are respectively characteristic density, temperature, mass fraction, pressure.

Nu and Sh are used for laminar flows or steady state flows. Stanton numbers, in most cases, characterize transport properties in a direction perpendicular to the flow [27] and are used for turbulent motions. The preceding dimensionless groups are generally defined together with other groups describing physical properties of the fluid.

b) *Characterization of fluid properties*

Heat, mass and momentum transfers are respectively characterized by thermal diffusivity coefficient D_T, diffusion coefficient D and kinematic viscosity coefficient, all expressed in $cm^2 \cdot s^{-1}$.

The fluid properties are characterized by the ratio of two of these quantities leading to:

The Prandtl number:

$$Pr = \frac{\nu}{a} = \frac{\mu c_p}{\varkappa} \text{ where } \nu \text{ is the kinematic viscosity} \tag{22}$$

The Schmidt number:

$$Sc = \frac{\nu}{D} = \frac{\mu}{\varrho D} \tag{23}$$

The Lewis number:

$$Le = \frac{D}{a} = \frac{\varrho C_p D}{\varkappa} \tag{24}$$

c) *Flow characterization*

Whether the fluid flow is laminar or turbulent can be predicted by the value of Reynolds number:

$$Re = \frac{uD}{\nu} \tag{25}$$

If the viscous stresses are sufficiently strong, instabilities will vanish rapidly and the flow remains laminar (Re is small). However at values of Re greater than a critical value (Re_c), instabilities are maintained and the motion is turbulent.

The analysis of experimental data, gathered for given geometrical and physical conditions, leads to empirical correlations for Nu and Sh, thus these dimensionless quantities can be expressed as function of Re, Pr and Sc.

$$Nu = f(Re, Pr) \tag{26}$$

and

$$Sh = f'(Re, Sc) \tag{27}$$

The main dimensionless group are listed in Table 2.

Table 2. Usual dimensionless numbers

Reynolds Number	$Re = \frac{Du\varrho}{\mu}$
Nusselt Number	$Nu = \frac{h_T D}{\varkappa}$
Prandtl Number	$Pr = \frac{d_p \mu}{\varkappa}$
Stanton Number	$St = \frac{h_T}{\varrho c_p u}$
Boltzman Number	$Bo = \frac{\varrho c_p u \pi}{n_0^2 T^3}$
Bringkham Number	$Br = \frac{\mu u^2}{\varkappa \Delta T}$
Eckert Number	$Ec = \frac{u^2}{c_p T}$
Lewis Number	$Le = \frac{\varrho c_p D_{ij}}{\varkappa}$
Schmidt Number	$Sc = \frac{u}{\varrho D_{ij}}$
Sherwood Number	$Sh = \frac{h_\omega D}{D_{ij}}$

2.3.1.4 Boundary Layer Approximation

Large Re numbers are equivalent to weak viscosity, as a consequence, the fluid may be considered as an ideal one and the velocity profile may be assumed to be flat [31]. However, such an approximation is not valid in the neighbourhood of the wall. For viscous flow fluid velocity must be zero at the wall, for ideal flow, only the normal velocity component must satisfy this condition. Thus, for large Reynolds numbers, velocity cancellation occurs in a thin layer, close to the rigid surface: the so called "dynamic boundary layer".

Rewritting (6) under the dimensionless formulation, the small quantity 1/Re appears on the right hand-side, and the contribution of viscous stresses may be considered as a small perturbation to an ideal fluid motion (in ideal flows right hand-side of (6) is zero). The boundary layer equations are obtained as velocity asymptotic expansions which must satisfy the perturbed equation [28] (Table 3). Physically, this is equivalent to say that the velocity gradient in the flow direction is very small compared with the normal one, and that the normal velocity component is much smaller than the axial one. It can be shown that the ratio of the transverse to longitudinal velocities is about $Re^{-1/2}$ [27,30] and that the boundary layer thickness varies as $Re^{-1/2}$ (Fig. 17). Such considerations may be applied to temperature and concentration profiles and lead to the so called "Thermal boundary layer" and "Diffusion boundary layer". According to similitude laws, Pr, Le, Sc numbers allows a comparison to be made of these different layer thicknesses:

Table 3. Boundary layer equations [28)]

— *Mass conservation*:

$$\frac{\partial}{\partial x}(\varrho_u r^k) + \frac{\partial}{\partial y}(\varrho v r^k) = 0 \qquad 3\text{-}1$$

— *x momentum component conservation*:

$$\varrho_u \frac{\partial u}{\partial x} + \varrho_v \frac{\partial v}{\partial y} = -\frac{dP}{dx} + \frac{\partial}{\partial y}\left(u \frac{\partial u}{\partial y}\right) \qquad 3\text{-}2$$

— *Species conservation*:

$$\varrho_u \frac{\partial Y_i}{\partial x} + \varrho_v \frac{\partial Y_i}{\partial y} = -\frac{\partial}{\partial y}(j_{D_i}) + \omega_i \qquad : i = 1 \ldots N \qquad 3\text{-}3$$

— *Specific internal energy conservation*:

$$\varrho_u \frac{\partial e}{\partial x} + \varrho_v \frac{\partial e}{\partial y} = -\frac{\partial q}{\partial y} - P\left(\frac{\partial u}{\partial x} + \frac{\partial v}{\partial y}\right) + u\left(\frac{\partial u}{\partial y}\right)^2 \qquad 3\text{-}4$$

k = 0 for plane flow, = 1 for rotationally symmetric flow, q is the energy and species diffusion flux as defined in Sect. 2.3.2.2.

Pr: dynamic/thermal boundary layers
Le: thermal/diffusion boundary layers
Sc: dynamic/diffusion boundary layers

2.3.2 Plasma-Wall Heat Transfer

2.3.2.1 Introduction

The main processes responsible for heat transfer are the following [37)]:

— in the boundary layer:
  — conduction and convection
  — convective diffusion

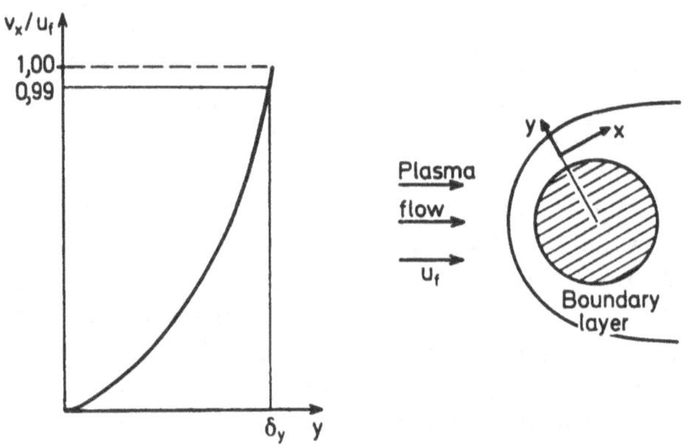

Fig. 17. Evolution of the dynamic boundary layer thickness

— to the wall: species recombinations
  — atom—atom→molecule
  — ions—e^-→neutrals

— radiative transfer.

The heat transfer phenomena and, in particular, the thermal boundary layer, are well understood for a cool, solid body immersed in a laminar hot gas stream in which no chemical reactions occur. The heat transfer can be predicted by the Nusselt number: Nu = f(Re, Pr). With some modifications, similar relationships hold for dissociation of gases, provided that Le, which describes the diffusion of the species, is close to unity [38].

If one considers heat transfer in a temperature range high enough for ionization to occur, one might be tempted to expect a strong increase of the heat transfer coefficient because the free electrons contribute strongly to the thermal conductivity, as they do in a metal.

One may consider two cases:

Firstly, walls are completely non-catalytic surfaces, and in this case, there is a high electronic concentration close to the wall leading to a strong thermal conductivity.

Secondly, the walls are fully catalytic. Recombinations maintain, close to the wall, a thin layer with a low density of charged species. Such a layer acts as a thermal insulator and prevents a large increase in heat transfer [39]. When this layer is under ionization equilibrium, its composition depends only on temperature, so it has the same thickness as the thermal layer.

If the ionized layer is "frozen", its composition is controlled by diffusion and the thickness is much smaller than in the equilibrium case (Fig. 18). Thus one sees the importance of Le number which characterizes the respective thicknesses of thermal and diffusion boundary layers. When Le increases, the diffusion layer becomes thicker and the heat transfer to the wall, decreases.

2.3.2.2 Transport Equations

Consider a gas mixture of two components, subscripted 1 and 2: respectively. The following equations are written in the boundary layer approximation, assuming that a concentration and temperature gradient exist only in the y direction (normal to the flow).

By the definition of the diffusion velocity in eq (3) one obtains [40]:

$$j_{D_1} = -\varrho D_{12}\left(\frac{\partial Y_1}{\partial y} + Y_1 Y_2 \alpha \frac{\partial}{\partial y}(\text{Log } T)\right) \tag{28}$$

where Y_1 and Y_2 are the mass fractions of the two components, ϱ is the density of the mixture, D_{12} the binary mass diffusion coefficient, α is the thermal diffusion ratio:

$$\alpha = \frac{\varrho}{\varrho_1 \varrho_2} \frac{D_{T,1}}{D_{12}}$$

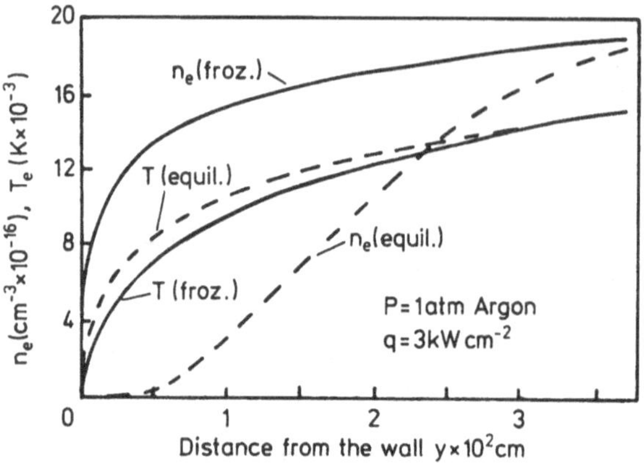

Fig. 18. Calculated electron density and temperature distribution in a plasma wall boundary layer. Reprinted from [38]

The heat flux becomes:

$$q = -\varkappa \frac{\partial T}{\partial y} + (\hat{h}_1 - \hat{h}_2)\, j_{D_1} - \alpha RT \frac{M^2}{M_1 M_2} j_{D_1} \tag{29}$$

The symbol $\varkappa$ denotes the thermal conductivity, $\hat{h}_1$ and $\hat{h}_2$ are the enthalpies of the components, R is the gas constant, M, M_1 and M_2 denotes the molecular weights of the mixtures and of the two components respectively.

Introducing j_{D_1} from eq. (28) into eq. (29) one obtains an analogous relation to eq. (10):

$$\begin{aligned} q = &-\left(\varkappa + \varrho D Y_1 Y_2 \left(\frac{\alpha}{T}(\hat{h}_1 - \hat{h}_2) - \alpha^2 \frac{RM^2}{M_1 M_2}\right)\right)\frac{\partial T}{\partial y} \\ &-\varrho D\left((\hat{h}_1 - \hat{h}_2) - \alpha RT \frac{M^2}{M_1 M_2}\right)\frac{\partial Y_1}{\partial y} \end{aligned} \tag{30}$$

giving the equivalent thermal conductivity:

$$\varkappa_e = \varkappa + \varrho D Y_1 Y_2 \left|\frac{\alpha}{T}(\hat{h}_1 - \hat{h}_2) - \alpha^2 \frac{RM^2}{M_1 M_2}\right| \tag{31}$$

In practice [38], the temperature gradient effect on diffusion is neglected and this gives for eq. (28) and eq. (29) respectively:

$$j_{D_1} = -\varrho D \frac{\partial Y_1}{\partial y} \tag{32}$$

$$q = -\frac{\mu}{Pr_f}\frac{\partial h}{\partial y} - \varrho D\left(1 - \frac{1}{Le_f}\right)(\hat{h}_1 - \hat{h}_2)\frac{\partial Y_1}{\partial y} \tag{33}$$

where Pr_f is the "frozen Prandtl number" defined as:

$$Pr_f = \frac{uC_{pf}}{\varkappa_f}$$

and where Le_f is the "frozen Lewis Number" defined as:

$$Le_f = \frac{\varrho C_{pf} D}{\varkappa_f}$$

C_{pf} denotes the frozen specific heat and $\varkappa_f$ denotes the frozen thermal conductivity. Thus Eq. (33) may be written as:

$$q = -\left(\frac{\mu}{Pr_f} + \varrho D\left(1 - \frac{1}{Le_f}\right)\frac{\partial Y_1}{\partial h}\;(\hat{h}_1 - \hat{h}_2)\right)\frac{\partial h}{\partial y}$$

Comparison with the first term of (33) on the right hand side leads to a definition of the "equilibrium" Prandtl number Pr_e as:

$$q = -\frac{\mu}{Pr_e}\frac{\partial \hat{h}}{\partial y}$$

The above equation allows the heat transfer to be described when chemical equilibrium is established. Note that when $Le_f = 1$ Eq. (33) and (34) are the same. This analysis, proposed by Pfender [38], has been verified using some experimental results which show that the frozen boundary layer hypothesis is correct. The measurements have been performed in a free burning argon arc with different water cooled probes. ($\varphi < 0.5$ mm). The influence of the charged particles has been determined by the measurements of the saturation current in the probe maintained at a floating potential.

2.3.2.3 *Empirical Correlations for Heat Transfer*

For a partially ionized gas, dimensional analysis leads to the following expression for the Nu number [40]:

$$Nu_f = f(Pr_f, Re, Sc, \xi, \varphi) \tag{35}$$

where ξ is the ionization degree and $\varphi = \dfrac{eV}{kT}$

Petrie [40] gives the following coefficients:

$$Nu_f = (0.064\,Pr_f^{1/3} + 0.0348\left(\frac{\xi\varphi^2}{Sc}\right)^m Re^{0.769} \tag{36}$$

where m is an exponent depending on the experimental conditions. The above expression represents the data within $\pm 15\%$ over the entire range of experimental conditions.

2.3.2.4 Case where the Walls are Electrodes

The extreme values of temperature, current density and field strength prevailing in the regions close to the electrodes and the gradients of these parameters make diagnostics a "formidable task" [38]. Nevertheless, the literature dealing with this subject is very extensive [41–44]. Heat transfer studies are complicated by three main difficulties.

Firstly, the conventional equations apply only as long as the continuum approach is valid. Since the anode fall spacing is of the order of one mean free path of the electrons the continuum approach is no longer valid for that part of the anode region.

Secondly, the application of conservation equations requires that the plasma is in LTE or, at least, that its thermodynamic state is known.

Finally, the specification of realistic boundary conditions for an anode region is extremely difficult.

The principal task of the anode fall region, namely to provide an electrical connection between the high temperature plasma of the arc and the low temperature anode, embraces several effects imposed by the conservation equations.

Conservation of energy requires that the field strength and current density adjust themselves so that the total net energy losses are compensated for by the ohmic heating. Conservation of charge carriers demands ion production in the anode fall region. At the surface, the current is exclusively carried by electrons which gives rise to a net negative space charge. The potential drop in the anode fall zone is a consequence of the net space charge adjacent to the anode surface (Poisson equation). It can be shown [44] that electrons entering the anode are responsible for the main part of heat transfer. The expression giving the heat flux going through the anode is:

$$q_a = q_c + q_r + j_e\left(\frac{5}{2}\frac{kT_e}{e} + V_a + q_a\right) \tag{37}$$

where q_c is the local heat flux by conduction and convection, q_r is the radiative flux from the plasma to the wall, j_e is the electron current density, T_e is the electron temperature, φ_a is the work function of the anode material and V_a is the anode fall.

In most cases heat transfer to the anode is governed by the electron flow:

$$q_a = j_e\left(\frac{5}{2}\frac{kT_e}{e} + V_A + \varphi_a\right)$$

2.3.2.5 Radiative Transfer Contribution

Generally the radiative contribution is much smaller than the convection contribution. However, the radiative flux increases strongly with temperature and pressure, and approaches that of black body radiation at P > 100 atm. Mostly one considers the Kramers radiation taking into account bremstrahlung processes and electron capture by heavy particles [45].

It can be shown [45] that in an optically thin gas one has:

$$\frac{\text{convective flux}}{\text{radiative flux}} = \frac{3}{16}\,\tau_0 B_\sigma \quad \text{for} \quad \tau_0 \gg 1$$

and in an optically thick gas:

$$\frac{\text{convective flux}}{\text{radiative flux}} = B_\sigma/2\tau_0 \quad \text{for} \quad \tau_0 \ll 1$$

B_σ is the Boltzman number defined by:

$$B_\sigma = \frac{\varrho u C_p}{\sigma T_0^3} \tag{38}$$

τ_0 is thermal boundary layer optical thickness.

2.3.2.6 *Correlations for Nusselt Number*

The different correlations are listed in Tables 4 and 5.

2.3.3 Plasma-Particles

A particle immersed in a fluid in motion is accelerated by viscous forces. Moreover, if the particle and fluid temperatures are different, heat transfer processes are occuring, generally accompanied by chemical reactions and mass transfer processes.

Table 4. Correlation for Nusselt Number

Nu	Conditions	Ref.
$1.86\left(\text{Re Pr}\,\frac{L}{D}\right)\left(\frac{\bar{u}}{u_f}\right)^{0.14}$	Re < 200; L/D > 10 Wall with T = constant	[47]
$0.026\,\text{Re}^{0.8}\,\text{Pr}^{1/3}\left(\frac{\bar{u}}{u_f}\right)^{0.14}$	Re > 210^4; 0.6 ≦ Pr < 100 . L/D > 10	[47]
$0.023\,\text{Re}^{0.8}\,\text{Pr}^{m}$	0.7 < Pr < 100; Re > 10^4 L/D > 60; m = 0.3 for cooling m = 0.4 for heating	[48]
$7 + 0.025\,(\text{Re Pr})^{0.8}$	Re · Pr > 100 Heat flux to the wall is constant	[49]
$5 + 0.025\,(\text{Re Pr})^{0.8}$	200 < Re · Pr < 10^4 T_{wall} = cte	[49]

Table 5. Correlation in the boundary layer

			Ref.
$Nu_{loc} = 0.323\ Re_{loc}^{1/2}\ Pr^{1/3}$	$Pr > 0.7$	Theoretical	50)
$Sh_{loc} = 0.332\ Re_{loc}^{1/2}\ Sc^{1/3}$	$Re_{loc} < 10^5$	results	
$Nu = 0.037\ Re^{0.8}\ Pr^{1/3}$	$Pr > 0.7;\ Re > 10^5$		51)
$Nu = 0.02\ Re_{loc}^{0.87}\ Pr^{0.33}$	Hot gas strinking on a wall		52)

2.3.3.1 Particle Motion

a) *Drag Coefficient*

Newton [53] presented experimental data for the force on a hollow sphere falling in air and he derived the following equation which seemed to predict his results theoretically:

$$F = \frac{\pi}{2} R_p^2 \varrho V^2 \tag{39}$$

According to dimensional analysis one introduces a dimensionless number used to characterize the drag force: this number is the so called "drag coefficient" which compares the drag force to the product of the dynamic pressure by the cross-sectional area of the particle, yielding:

$$C_D = \frac{F}{\frac{1}{2} \varrho u_r^2 A} \tag{40}$$

The expression giving the C_D value is generally valid for restrictive conditions characterizing fluid motion around the particle. For that reason the drag coefficient is expressed as a Re function.

The first results are due to Stokes [54] theoretical analysis leading to the expression:

$$C_{D_S} = \frac{24}{Re} \tag{41}$$

Prupacher [55] proposed a representation of the deviation to Stokes' law by the formula:

$$\frac{C_D}{C_{D_S}} - 1 = \alpha\ Re^{\beta} \tag{42}$$

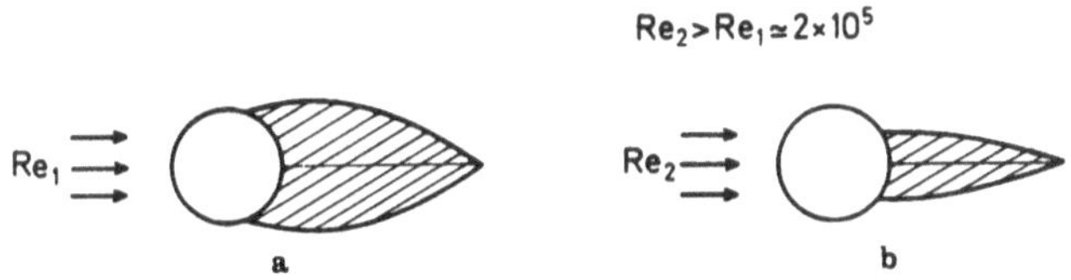

Fig. 19. Turbulent wake dimensions behind a particle

and gave the following relations:

$$\begin{aligned}\frac{C_D}{C_{D_S}} - 1 &= 0.102\ \mathrm{Re}^{0.95} && 10^{-3} < \mathrm{Re} < 2\\ &= 0.115\ \mathrm{Re}^{0.80} && 2 < \mathrm{Re} < 21\\ &= 0.189\ \mathrm{Re}^{0.632} && 21 < \mathrm{Re} < 200\end{aligned} \tag{43}$$

Stokes' analysis assumes that the flow about the sphere is symmetrical. This assumption is valid only for low Re numbers (Re < 0.05). For higher Re values the flow presents a turbulent wake downstream the particle (Fig. 19).

As the Re value increases, C_D decreases and takes a minimum value (0.38) at Re = 5000; then C_D rises slowly up to 0.5 for Re = 70000. In the Newton region (1000 < Re < 200000), C_D is approximately constant (~0.5). Clift and Gauvin [56] give the relation:

$$C_D' = 24\ \mathrm{Re}^{-1}(1 + 0.15\ \mathrm{Re}^{0.687}) + 0.42(1 + 4.25 \cdot 10^4\ \mathrm{Re}^{-1.16})^{-1} \tag{44}$$

For Re ~ 300000 the "resistance crisis" occurs. The drag force depends on wake dimensions and these dimensions also depend upon the position of the separation line between the wake and the boundary layer. As Re increases, the turbulent separation point moves back; as a consequence the wake dimensions diminish and the drag coefficient drops sharply from 0.4 to 0.1 (Fig. 20). All of these considerations are reported in detail by Torobin and Gauvin [57]. Boulos [58], Bhattacharyya and

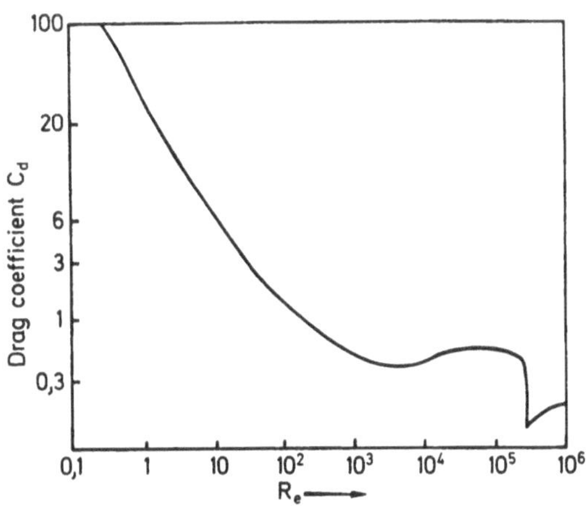

Fig. 20. Standard drag curve

Table 6. Correlations for the drag coefficient [59)]

Re	C_D	
	Spherical particle	Flakes: e 0.026 D_p
$Re < 0.1$	$24/Re$	$24\,S_1/Re$
$0.1 \leqq Re < 1$	$(24/Re)(1 + 3\,Re/16)$	$(24\,S_1/Re)(1 + 3\,S_1\,Re/16)$
$1 \geqq Re < 20$	$(24/Re)(1 + 0.11\,Re^{0.810})$	$(20.4/Re)(1 + 0.136\,Re^{0.803})$
$20 \geqq Re < 200$	$(24/Re)(1 + 0.189\,Re^{0.632})$	$(20.4/Re)(1 + 0.138\,Re^{0.793})$

S_1 is a shape factor; for thin disks $S_1 = 8/3\pi$

Gauvin [59)] suggest the correlations given in Table 6. Moreover, in order to consider the important viscosity variations in the boundary layer Lewis and Gauvin [60)] adopt the following correction:

$$C_D = \bar{C}_D \left(\frac{\bar{\nu}}{\nu f}\right)^{0.15}$$

b) *Particle Trajectories*

Particle trajectories are obtained by solving the "Basset-Boussinesq-Oseen" equation [58, 61)]:

$$\frac{\vec{F}}{\pi D_p^2} = -\frac{C_D \varrho_f}{8} |\vec{U}_r| \vec{U}_r - C_A \varrho_f \frac{D_p}{6} \frac{d\vec{U}_r}{dt} - C_H \sqrt{\varrho_f \frac{\mu_f}{6}} \int_0^t \left(\frac{d\vec{U}_r}{dt}\right)_{t=\varphi} \cdot \frac{d\varphi}{\sqrt{t-\varphi}} \tag{46}$$

$\vec{F}$ is the force acting on the particle of diameter D_p. The first term in R.H.S. is the drag force contribution, $\vec{U}_r$ is the relative velocity between the particle and the fluid.

The second term is the added mass term where C_A is the added mass coefficient. The last term in Eq. (46) is the Basset history term where C_H is the history coefficient, ϱ_f is the density of the fluid, μ_f the viscosity of the fluid and φ is a dummy parameter. In most cases the added mass and the history terms are orders of magnitude less than the drag term and, consequently, may be neglected leading to the equation:

$$\frac{d\vec{U}_p}{dt} = -\frac{3}{4} C_D |\vec{U}_p - \vec{U}_f| (\vec{U}_p - \vec{U}_f) \frac{\varrho_f}{\varrho_p D_p} \tag{47}$$

where $\vec{U}_p$ is the particle velocity, $\vec{U}_p$ the fluid velocity and ϱ_p is the particle density.

When the flow presents axial, radial and angular velocity components, the force balance gives the following equations for the three directions

$$\frac{dU_{p_z}}{dt} = \frac{\varrho_f}{\varrho_p} \frac{dU_{f_z}}{dt} - \frac{3}{4} C_D \frac{\varrho_f}{\varrho_p D_p} |U_{p_z} - U_{f_z}| (U_{p_z} - U_{f_z}) + g\left(1 - \frac{\varrho_f}{\varrho_p}\right) \tag{48-a}$$

$$\frac{dU_{p_r}}{dt} = \frac{\varrho_f}{\varrho_p}\frac{dU_{f_r}}{dt} - \frac{3}{4}C_D\frac{\varrho_f}{\varrho_p D_p}|U_{p_r} - U_{f_r}|(U_{p_r} - U_{f_r})$$
$$+ 2\left(1 - \frac{\varrho_f}{\varrho_p}\frac{U_{f_r}^2}{U_{p_\varphi}^2}\right)\frac{U_{p_\varphi}^2}{D_p} + \sqrt{\nu_f K}\,(U_{p_\varphi} - U_{f_\varphi})\,81\varrho_f/(2D_p\varrho_p) \quad (48\text{-b})$$

$$\frac{dU_{p_\varphi}}{dt} = \frac{\varrho_f}{\varrho_p}\frac{dU_{f_\varphi}}{dt} - \frac{3}{4}C_D\frac{\varrho_p}{\varrho_p D_p}|U_{p_\varphi} - U_{f_\varphi}|(U_{p_\varphi} - U_{f_\varphi})$$
$$- 2\left(1 - \frac{\varrho_f}{\varrho_p}\frac{U_{f_\varphi}^2}{U_p U_{p_r}}\right)\frac{U_{p_\varphi}U_{p_r}}{D_p} \quad (48\text{-c})$$

Here, subscripts p and f denote the particle and the fluid respectively, subscripts z, r, φ denote the axial, radial and angular direction respectively. In the three equations the first term in R.H.S. is the buroyancy force and the second term is the drag force. The last terms in Eq. (48-a), Eq. (48-b) and Eq. (48-c) are gravity, centrifugal force and Coriolis force respectively. K appearing in Eq. (48-b) is the curl of the fluid velocity, supposed to be axially oriented.

2.3.2.2 Plasma-Particle Heat Transfer

a) *Heat balance equations*

If one first considers that the temperature increases up to the melting point the heat flux received by the particle is [61, 62]:

$$q = h_T S_p(T_g - T_p) - h_r(T_p) =$$
$$= h_T S_p(T_g - T_p) - S_p\varepsilon\sigma(T_p^4 - T_0^4) \quad (49)$$

where h_T and h_r are respectively the convective and radiative coefficients. S_p the particle area, m_p, C_p are the mass and the specific heat of the particle. ε is the emissivity, σ the Stefan constant, and T_g, T_p and T_0 are the temperature of the gas, particle and of the reactor wall, respectively.

The first and second terms in the right hand side are respectively the convective and radiative contributions (plasma radiative contribution is generally neglected). One part of the energy is absorbed by specific heat and the other part is lost by radiative transfer giving:

$$m_p C_p \frac{dT_p}{dt} = h_T S_p(T_g - T_p) - S_p\varepsilon\sigma(T_p^4 - T_0^4) \quad (50)$$

When the melting point is reached, the temperature is assumed to be constant and the heat flux provided to the particle increases the liquid fraction x, according to the expression [56]:

$$mL_m \frac{dx}{dt} = h_T S_p(T_g - T_p) - S_p\varepsilon\sigma(T_p^4 - T_0^4) \quad (51)$$

Finally, after complete melting, the temperature rises again until it reaches the boiling point. Then the particle temperature remains constant and its diameter decreases by the evaporation processes. One has [61]:

$$\frac{\varrho_p \pi D_p^2}{2L_v} \frac{dD_p}{dt} = h_T S_p(T_g - T_p) - S_p \varepsilon \sigma (T_p^4 - T_0^4) \tag{52}$$

This equation stipulates that no mass losses are occuring before reaching the boiling point. Generally evaporation may occur before this step and, moreover, some materials may not pass through the liquid phase.

Two evaporation processes may be postulated [63]:
firstly the plasma is inert and the most important reactions are those of decomposition: $AB \rightleftarrows A + B$; A, B are gaseous or condensed elements or compounds; secondly the plasma is reactive and heterogeneous phase reactions are involved according to:

$$A(g) + B \rightleftarrows C + D(g)$$

A more sophisticated model involving diffusion process proposed by Bonet [63] is described in the next section.

b) *Reactive model*

The gaseous phase issuing from the material tends to maintain a vaporisation-condensation equilibrium. Let ϱ_{vs} be the mass concentration of the vapour at the particle surface and ϱ_v the same quantity away from the surface. The mass evaporation flux is given by:

$$j^v = h_\omega(\varrho_{vs} - \varrho_v) \tag{53}$$

where h_ω is defined in (15).

The evaporation flux induces the particle radius variation according to:

$$\frac{dR_p}{dt} = \frac{j_v}{\cdot\ \varrho_p} \tag{54}$$

This flux gives rise to the corresponding enthalpy flux: $j_v \Delta H_t$ with:

$$\Delta H_t = \Delta H_{reac} + (h_v(T_g) - h_g(T_g) - (h_v(T_p) - h_g(T_p)) \tag{55}$$

where ΔH_{react} is the enthalpy of the reaction occuring at the surface, h_v is the mass enthalpy of the gaseous product, and h_g is the mass enthalpy of the plasma. Thus (50) must be rewritten:

$$h_T S_p(T_g - T_p) - S_p \varepsilon \sigma (T_p^4 - T_0^4) = mC_p \frac{dT_p}{dt} - j_v S_p \Delta H_T \tag{56}$$

One may consider the material evaporation by introducing the characteristic time leading to a dR_p radius variation. If the process is controlled only by diffusion one has the dt_D time as:

$$dt_D = \varrho_p \frac{dR_p}{j_v} \tag{57}$$

Expressing h_ω as a Sh number function and noting that $\varrho_{vs} \gg \varrho_v$ Eqs. (57) and (53) lead to:

$$dt_D = 2 \frac{\varrho_p}{\varrho_v} \frac{R_p}{\text{Sh } D\varrho_{vr}} dR_p \tag{58}$$

If now one considers that heat transfer controls the process j_v must be expressed by Eq. (56). Neglecting radiative contributions and expressing h_T as Nu function Eq. (57) gives the heat transfer dt_T time:

$$dt_T = 2\varrho_p \frac{\Delta H_T R_p}{\text{Nu } \varkappa(T_g - T_p)} dR_p \tag{59}$$

Thus one has two consecutive and competitive processes (before diffusing, the vapour must be produced by heat transfer). The material evaporation is controlled by the slowest of the two processes. Expressing the dt_D/dt_T ratio one has:

$$\frac{dt_D}{dt_T} = \frac{\text{Nu}}{\text{Sh}} \frac{\varkappa}{\varrho_v} \frac{(T_g - T_p)}{\Delta H_T D\varrho_{vs}} \tag{60}$$

The heat and mass transfer analogy allows to write: Nu/Sh = 1 (from (27) this implies that Pr = Sc). If one supposes that Le is close to 1, that is:

$$\text{Le} = \varkappa/(C_p\varrho_v D) \sim 1$$

one finally obtains:

$$\frac{dt_D}{dt_T} = \frac{C_p(T_g - T_p)}{\varrho_{vs} \Delta H_T} \tag{61}$$

Usually $\Delta H_T > C_p(T_g - T_p)$. If ϱ_{vs} is low then $dt_D > dt_T$; the process is controlled by the diffusion. If ϱ_{vs} is important then $dt_D < dt_T$; the decomposition process can be controlled by heat transfer.

The radiative transfer, contribution which cannot be neglected as particles ($\varphi < 200\,\mu$) reach the melting point, tends to increase dt_T, that is, reinforces the tendency for heat transfer to control the process.

c) *Influence of the particle thermal conductivity*

A model in which temperature gradient is involved, has been developed by

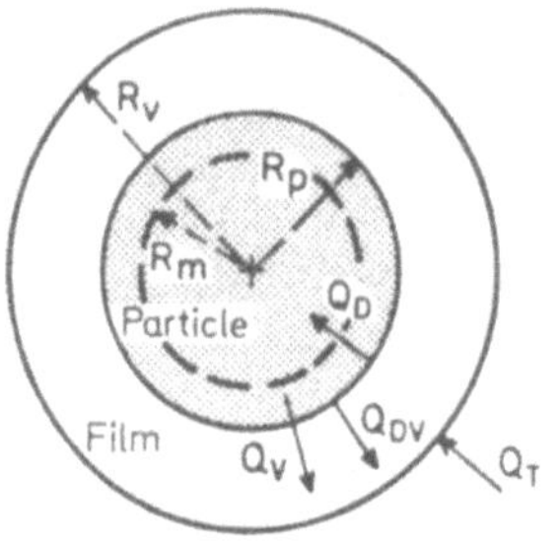

Fig. 21. Schematic diagram of heat transfer to vapor film and liquid particle. Reprinted from [64]

Yoshida et al. [64] and Lesinski et al. [65]. When an evaporation flux leaves a partially melted particle, the energy balance may be written as [64]:

$$Q_T = Q_D + Q_V + Q_{DV} \tag{62}$$

where Q_T is the total energy provided by the plasma to the particle,
Q_D is the energy provided to the solid core,
Q_V is the energy required for evaporation and
Q_{DV} is the energy required for heating up the gaseous products.

The transfer mechanism is shown in Fig. 21.

The differential equation for temperature in a spherical particle is:

$$\varrho_p C_p \frac{\partial T}{\partial t} = \frac{1}{r^2} \frac{\partial}{\partial r} \left(r^2 \varkappa \frac{\partial T}{\partial r} \right) \tag{63}$$

The boundary conditions required for the resolution depend on temperature distribution in the particle. Several cases must be distinguished.

i) The particle is not melted (Fig. 22a). At the external surface this condition can be expressed in the following form:

$$\varkappa_{ps} \left(\frac{\partial T}{\partial r} \right)_{r=R_p} = h_T(T_g - T_p) \tag{64}$$

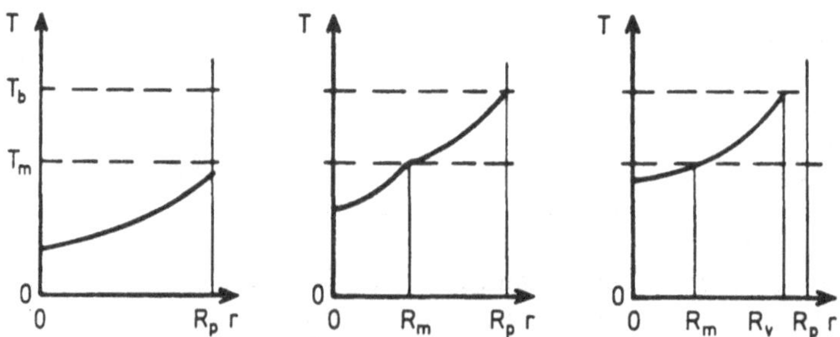

Fig. 22a—c. Radial temperature distribution in an alumina particle:
a melted particle, **b** partially melted particle, **c** evaporation. Reprinted from [65]

The heat flux entering the particle through the external surface is only required for heating up the solid material until the surface temperature reaches the melting point. The particle then begins to melt and a moving front results which changes the boundary conditions.

ii) The particle is partially melted (Fig. 22b). In addition to an expression like (64) holding at the external boundary, it is necessary to write another expression involving the melting front propagation. The heat balance expressed at the moving interface gives:

$$\varkappa_{ps}\left(\frac{\partial T}{\partial r}\right)_{R_m-\varepsilon} - \varkappa_{pl}\left(\frac{\partial T}{\partial r}\right)_{R_m+\varepsilon} = L_m \varrho_p \left(\frac{dR_m}{dt}\right) \tag{65}$$

$\varkappa_p$ is the thermal conductivity of particle material, the subscripts s and l denote the solid phase and the liquid phase respectively. The temperature gradient $\frac{\partial T}{\partial r}$ is evaluated in the neighbourghood of the melting front, the radius of which is R_m, at $r = R_m - \varepsilon$ for the solid phase and at $r = R_m + \varepsilon$ for the liquid phase, ε being an infinitely small quantity. Of course temperature continuity must be expressed:

$$T(R_m) = T_m \tag{66}$$

One part of the heat flux reaching the melting front is required for its propagation, the other part goes for heating up the solid core.

iii) Occurence of evaporation (Fig. 22c). As the surface temperature is increasing, the vapour pressure of the material increases too, giving rise to an evaporation flux. This flux requires a part of the energy provided by the plasma to the particle leading to the condition at the vaporizing front:

$$h_T(T_g - T(R_v)) = \varkappa_{pl}\left(\frac{\partial T}{\partial r}\right)_{r=R_v} + J_v\, \Delta H_T \tag{67}$$

Equation (67) must be applied at the liquid-gas interface, the radius variation being given by Eq. (54). Equation (67) may be written in the following form:

$$\varkappa_{pl}\left(\frac{\partial T}{\partial r}\right)_{r=R_v} = h_T(T_g - T(R_v)) - j_v\, \Delta H_T \tag{68}$$

If the right hand side is positive, the particle temperature progressively increases. If it is zero, the temperature remains constant and a stationary regime is reached.

Now, consider the term $j_v \Delta H_T$. The mass flux depends mainly on the vapour pressure, that is on surface temperature, and on the mass transfer coefficient h_ω. Moreover, as a consequence of the high plasma temperature gradient, the $h_T(T_g - T_p)$ term rapidly decreases as the particle continues on its trajectory. Though T_g is much higher than T_p, it may happen that the right hand side of Eq. (68) becomes negative, as the first term is decreasing more rapidly than the second one. In that case the $\left[\left(\frac{\partial T}{\partial r}\right)_r = R_v\right]$ quantity is negative and this condition means that

Table 7. $\varkappa_g/\varkappa_p$ conductivities ratio [66]

Powder material	Thermal conductivity of powder material $W \cdot m^{-1} \cdot K^{-1}$	$\varkappa_g/\varkappa_p$ for different plasma gases			
		Ar	N_2	NH_3	H_2
ZrO_2	2.39	0.18	0.63	3.5	5.5
Al_2O_3	5.86	0.07	0.26	1.4	2.2
MgO	5.86	0.07	0.26	1.4	2.2
BN	9.2	0.05	0.16	0.9	1.4
SiC	41.90	0.01	0.04	0.2	0.3
Ti	18.00	0.02	0.08	0.5	0.7
Fe	40.00	0.01	0.04	0.2	0.3
W	100.00	0.004	0.015	0.08	0.1

Comments: Nu = 2; the mean plasma temperatures are: 10000 K; N_2: 5000 K; H_2 and NH_3: 3500 K. The thermal conductivity of powder material is calculated for $T = 0.5\,T_{melt}$

the particle temperature is above the equilibrium temperature. A portion of the particle internal energy must therefore be used to supply the heat of vaporization. As a consequence, the particle temperature decreases until the balance of heat and mass transfer is satisfied [64] (see Section 2.3.4.2.c).

From Eq. (67) it can be seen that two processes are competitive as soon as evaporation takes place. (The gaseous products are coming from vaporization, from decomposition or from both processes.)

To favour particle melting, it is necessary for heat to propagate easily through the material. Thus $\varkappa_p$ must be large. In that case the fusion process is controlled by energy supply, that is, by the thermal conductivity of the plasma. On the other hand, if $\varkappa_p$ is low the fusion process may be controlled by the material conductivity $\varkappa_g/\varkappa_p > 1$. Rykalin [66] gives some characteristic values for different materials and plasma gases (see Table 7).

In the case of melting or vaporizing, moving-boundary problems must be solved; Yoshida et al. [64] adopted the variable space grid method which was developed by Murray and Landis [67].

d) *Transfer parameters*

These parameters are the Nu and Sh numbers. The usual expressions are [68]:

$$\begin{aligned} Nu &= 2 + 0.6\,Pr^{1/3}\,Re^{1/2} \\ Sh &= 2 + 0.6\,Sc^{1/3}\,Re^{1/2} \end{aligned} \tag{69}$$

The heat transfer coefficient h_T is obtained from Nu according to the equation:

$$h_T = \frac{\varkappa\,Nu}{D_p}$$

Calculating $\varkappa$ for a mean temperature value leads generally to a rather poor result. It is better to use the value of $\bar{\varkappa}$ defined by the relation:

$$\bar{\varkappa} = \frac{1}{T_g - T_p} \int_{T_p}^{T_g} \varkappa(T)\, dT \tag{70}$$

The Re and Pr values may therefore be obtained for a mean temperature value. Moreover, according to the viscosity variation over the boundary layer, the following modified Nu value is generally used [55]:

$$Nu = \overline{Nu} \left(\frac{\bar{\nu}}{\nu_f}\right)^{0.15} \tag{71}$$

When decomposition reactions or vaporization occur, the gaseous products at the particle surface tend to modify the value and the following correction has been proposed [57]:

$$h_T' = h_T \left(\frac{1+\xi}{\xi}\right) \mathrm{Log}\,(1+\xi) \tag{72}$$

with:

$$\xi = \frac{\Delta H_T - \Delta H_{re}}{\Delta H_{re}} \tag{73}$$

where ΔH_T and ΔH_{re} are defined by expression (55)

2.3.4 Examples of Calculations and Measurements

2.3.4.1 The Plasma

a) *Influence of the nature of the gas on flow velocity*

From a qualitative point of view one can say that viscous stresses tend to determine the flow in a channel. If we now consider argon, nitrogen and hydrogen, which are the most frequently used gases in plasmas, typical working conditions (electric power and gas flow rates) for a DC plasma arc generator lead to the following mean temperature (corresponding to an homogeneous flow for which enthalpy is known and is supposed to be in equilibrium: enthalpy temperature): $T_{Ar} = 10000$ K, $T_{N_2} = 6500$ K, $T_{H_2} = 3500$ K. The viscosities respectively calculated for the above temperature are: $3.0 \cdot 10^{-3}$ g/cm/s for Ar, $1.6 \cdot 10^{-3}$ g/cm/s for nitrogen and $0.4 \cdot 10^{-3}$ g/cm/s for hydrogen. Thus, the respective flow velocity are $U_{H_2} > U_{N_2} > U_{Ar}$.

b) *Gas velocities and temperature in RF plasma*

Temperature and velocity distribution in a RF plasma generator (Fig. 23a) have been determined by Boulos [61] by solving the balance equations for a fluid submitted to electric and magnetic fields. The details of the calculations are reported in [69].

$r_1/R_o = 0.06$
$r_o/R_o = 0.20$
$R_1/R_o = 0.71$
$R_c/R_o = 1.71$
$L_o/R_o = 1.0$
$L_1/R_o = 4.0$
$L_2/R_o = 6.2$
$L_T/R_o = 13.0$

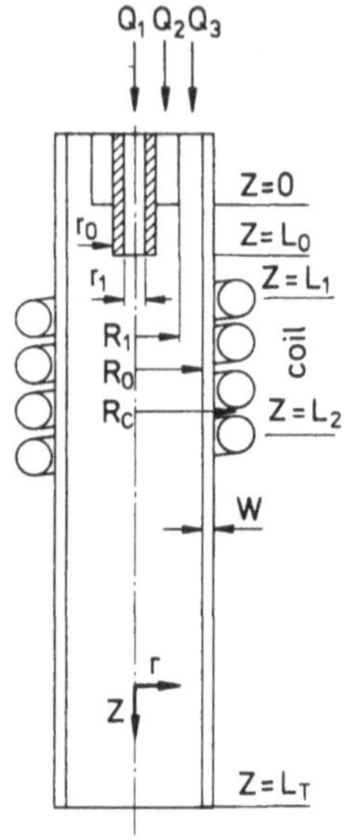

$r_1 = 0.168$ cm
$r_o = 0.28$ cm
$R_1 = 1.0$ cm
$R_o = 1.4$ cm
$R_c = 2.4$ cm
$W = 0.1$ cm
$L_o = 1.4$ cm
$L_1 = 5.6$ cm
$L_2 = 8.7$ cm
$L_T = 18.2$ cm

a

Torch geometry and system of coordinates

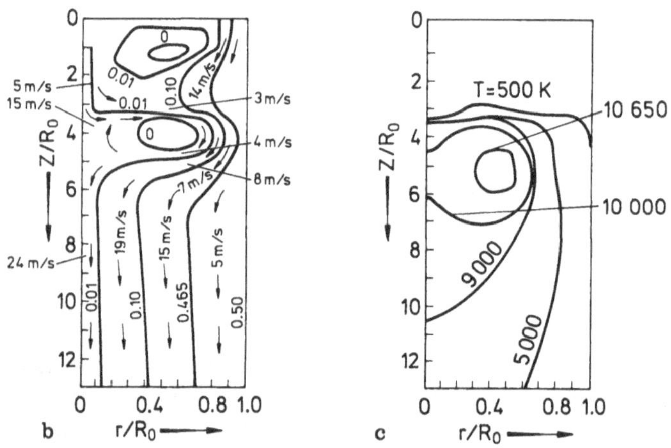

b, c Flow and temperature fields for Ql = 0.4 l/min

Fig. 23 a. Torch geometry and system of coordinates, **b** Stream lines and **c** temperature fields in a RF plasma torch working with argon. Reprinted from [61]

The stream lines and temperature contours are presented in Fig. 23b and 23c respectively. The calculations were performed for a frequency of 3 MHz and 3.77 kW electric power. The central gas flow rate, Q2, and the sheath gas flow rate, Q3, were kept constant at 0.2 and 1.6 l/mn respectively. Calculations were made with different particle carrier gas flow rates, QL_1 over the range 0 to 0.6 l/mn.

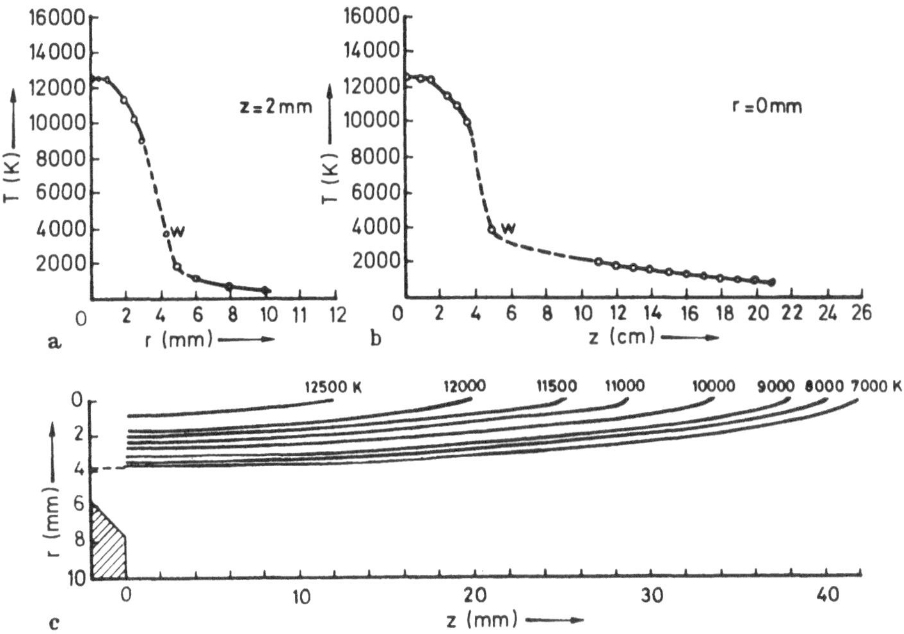

Fig. 24. Temperature distribution in a DC argon-hydrogen plasma jet. Reprinted from [65)]

c) *Gas velocities and temperature in a DC plasma*

Figures 24 and 25 show experimental results obtained in an argon-hydrogen plasma [65)]. The experimental conditions were: arc voltage: 52 V; current intensity: 500 A; argon flow rate: 44 l/min; hydrogen flow rate: 7 l/min.

Temperatures higher than 8000 K were measured by spectroscopic techniques. For lower temperatures the results have been obtained by thermocouple measurements ($T < 2000$ K) and by using the tungsten melting point. Figure 24a gives the radial evolution of the temperature 2 mm downstream the nozzle exit. Figure 24b presents the axial evolution of the jet temperature and Fig. 24c shows spatial temperature distribution in the free burning jet.

Gas velocities have been calculated from dynamic pressure, measured with a Pitot tube without cooling, the probe crossing the jet, for a given axial position, with an oscillating motion. The signal provided by capacitive coupling was analysed on an oscilloscope. The axial velosity distribution is shown in Fig. 25a and the spatial distribution in the free burning jet is given in Fig. 25b.

2.3.4.2 The Particles

a) *Particle trajectories*

In RF plasma

Boulos [61)] introduces in his model the magnetic pumping effect. Moreover he shows the influence of particle diameter and of the initial radial position of the particles in the powder feed tube on the trajectories. The reactor is shown in Fig. 23a. The modelling has been realized for alumina particles with 10, 30, 60, 100, 150, 200

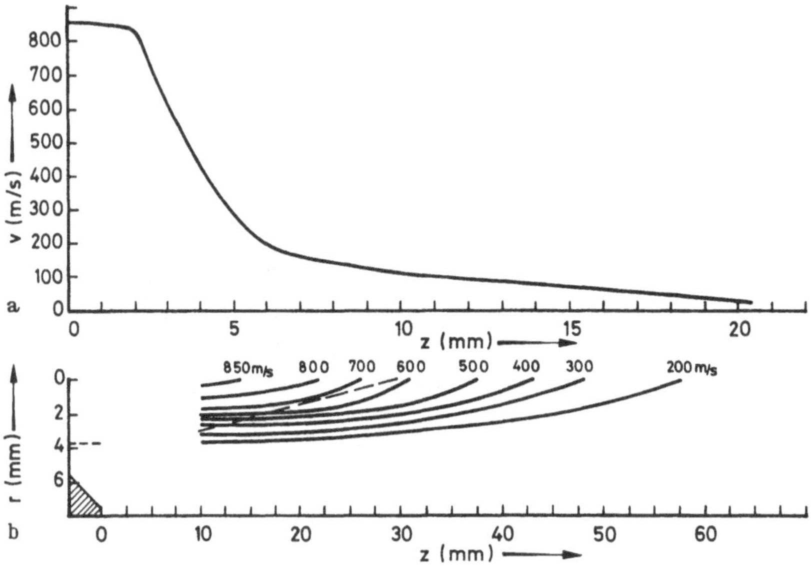

Fig. 25. Velocity distribution in a DC argon-hydrogen plasma jet. Reprinted from [65]

and 250 nm diameters. The initial radial position r_1 is defined in the figures by the (r_1/R_0) ratio. This ratio takes the following values: 0.01; 0.03; 0.05.

Figure 26a shows the trajectory of 10 µm particles which tend to follow the stream lines of the flow. As they enter the torch, they move vertically downwards until they reach the fire ball. At this point they just skim over its surface and move radially outwards where they are entrained by the sheath gas flow in the space between the fire ball and the quartz tube. Being so small, they are drawn into the fire ball by electromagnetic pumping. As the particles are exposed to the high temperatures encountered they vaporize very quickly. (Note that an open circle indicates a solid particle, a shaded circle represents a liquid droplet and a partially shaded circle stands for a partially melted particle.)

The larger 30 µm particles are not drawn into the fire-ball by the magnetic pumping (Fig 26b). They simply move with the sheath gas where they solidify. The 100 µm particles injected into the center move along the reactor axis and are completely melted after passing through the hottest zone (Fig. 26c). If the injection axis is radially displaced ($r_1/R_0 = 0.05$) the particles bounce off the fire ball, bounce again on the wall and finally enter the hot zone where they are melted Fig. 26d.

Boulos observed that 200 and 250 µm particles injected at $(r_1/R_0) = 0.01$ and 0.03 passed straight through the fire ball, while those injected at $r_1/R_0 = 0.05$ bounced vigorously all over the torch as shown in Fig. 26e and f. The 250 µm particles, entrained by their inertia, enter the fire ball neighbourhood where they are partially melted.

In this case the particle velocities are considerablely higher than those in a RF plasma. Generally, the particles are radially injected and the injection velocity and

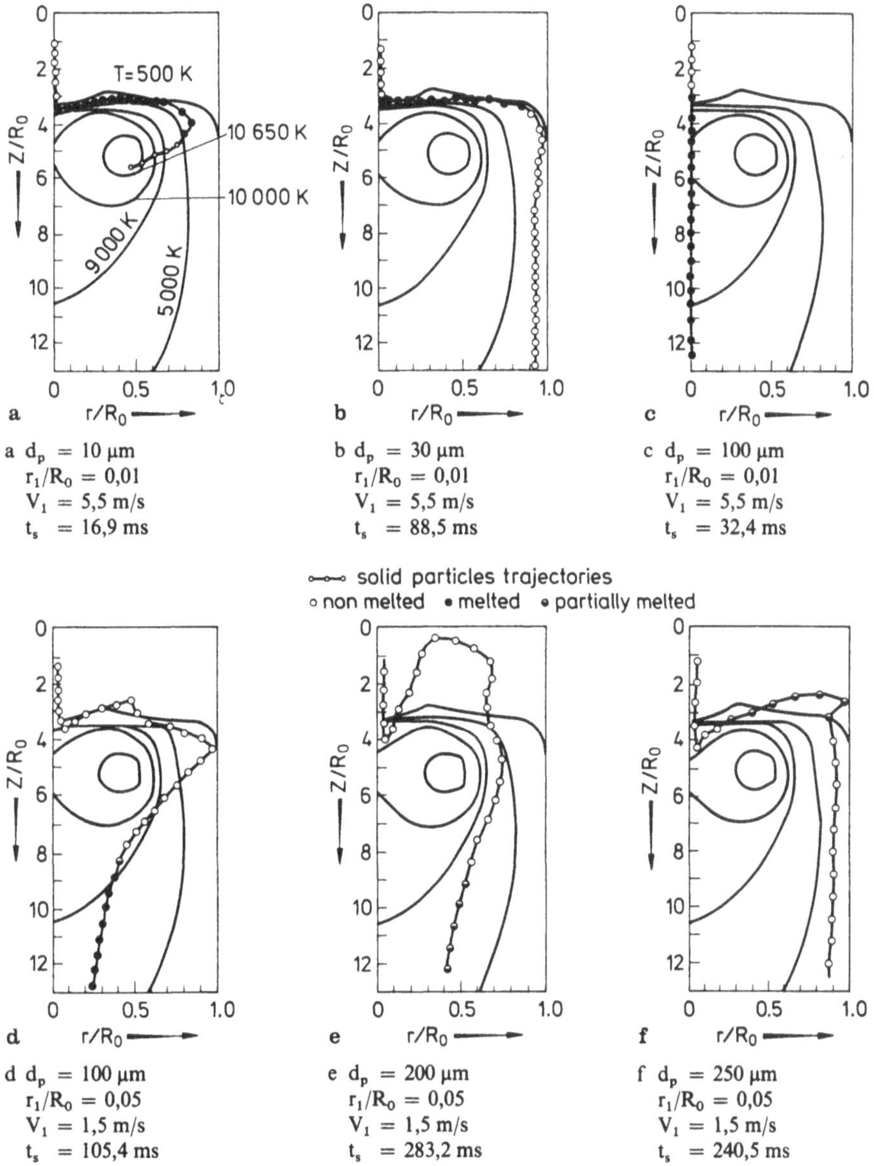

Fig. 26. Alumina particle trajectories in an RF argon-plasma. Reprinted from [61]

angle are important parameters. Boulos and Gauvin [58] demonstrate these influence on molybdenum disulfide particle trajectories as shown in Fig. 27.

b) *Particle temperature*

Iron particles in a RF plasma [64].

The axial temperature and velocity distributions used in this work are those calculated by Boulos [70] for an argon RF plasma and represented in Fig. 28a.

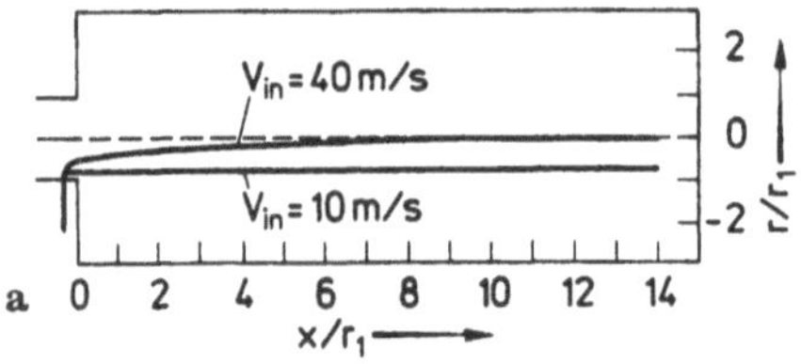

— Single particle trajectories in a d.c. plasma jet at 90° injection
$r_1 = 0.5$ cm, $u_1 = 423$ m/s, $t_1 = 12000$ K, $D_p = 30$ μm

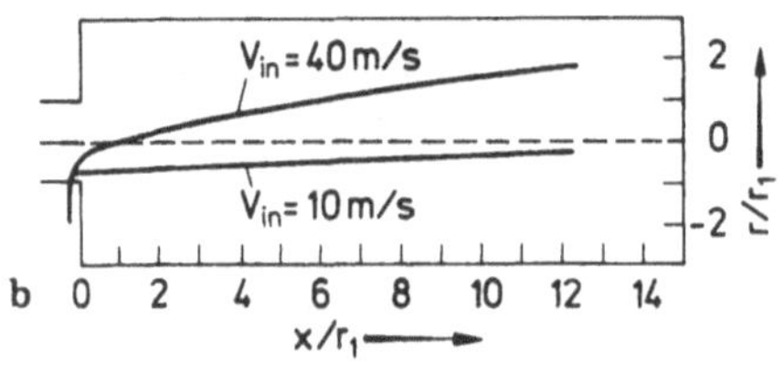

— Single particle trajectories in a d.c. plasma jet at 90° injection
$r_1 = 0.5$ cm, $u_1 = 423$ m/s, $t_1 = 12000$ K, $D_p = 60$ μm

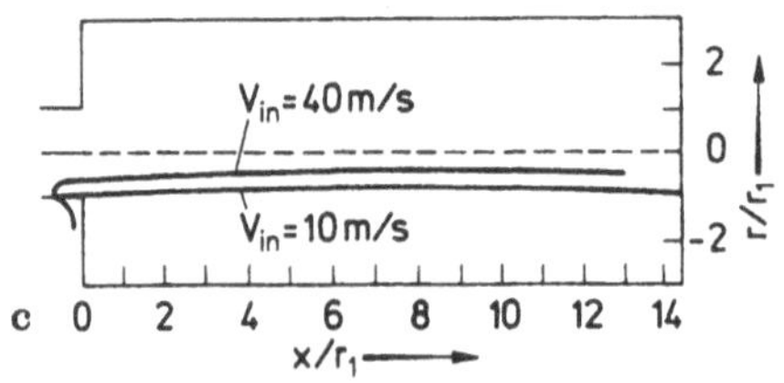

Fig. 27a—c. MoS_4 particle trajectories in a DC plasma jet. Reprinted from [58]

— Single particle trajectories in a d.c. plasma jet at 150° injection
$r_1 = 0.5$ cm, $u_1 = 423$ m/s, $t_1 = 12000$ K, $D_p = 30$ μm

The model is applied to iron particles with diameters of 20, 40, 60, 80 μm. The axial velocities of the particles A, B, C, D are given in Fig. 28b. Figure 28c shows the temperature history of each particle. The open and shaded circles represent their axial position 0.01 s and 0.02 s respectively after injection. All the curves show a change of slope near the Curie temperature. It is caused by the fact that the thermal diffusivity curve has a minimum value (singular point) at the Curie temperature. The A, B, and C particles are completely melted after 5.4, 10.1 and 15.0 m/sec following injection. The D particles reach only 1759 K and are not melted. The melting conditions for A, B and C particles are summarized in Table 8.

From Fig. 28c note that the surface temperature of particle C increases gradually until it reaches the flame temperature. Particles A and B approached asymptotically to about the same temperature, that is 0.93 of the boiling temperature. Note that their surface temperature decreases even though the plasma temperature is much higher. This results from considerations mentioned in Section 2.3.3.2.c. The main results of this simulation are listed in Table 9.

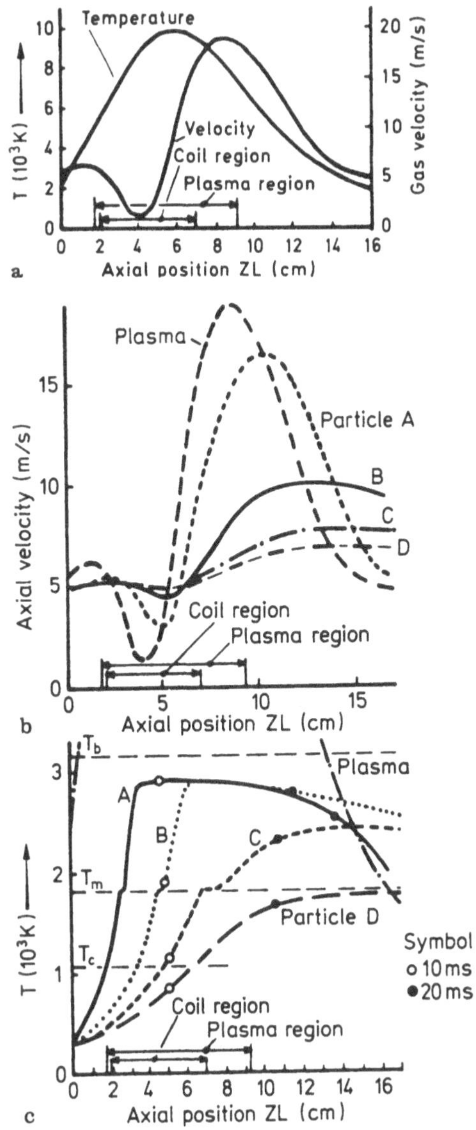

Fig. 28 a, b, c. Predicted iron particle surface temperature and velocity in a RF argon plasma. Reprinted from [64]

Alumina particles in a DC plasma jet [65]

The temperature and velocity distributions in the plasma jet are those mentioned in Section 2.3.4.1.c. The particles are 40 μm in diameters and are assumed to move along the jet axis.

Fusion is almost complete 7 mm from the injection region, the corresponding residence time being about 0.14 m sec. During this time, the surface reaches the boiling temperature (Fig. 29 a). The core reaches the same temperature 30 mm downstream of the nozzle exit (0.23 m sec) and at this step the particle radius is about 50 % smaller than the initial radius as a consequence of evaporation.

Table 8. Melting conditions of iron particles in an argon RF plasma (after [64])

Particle	Plasma Temperature (K)	Axial position (cm)	Melting time (m sec)	Plasma velocity (m/sec)	Particle velocity (m/sec)	Heat transfer coefficient 10^4 W/m²/K)
A (20 μ)	6410–6760	2.54–2.74	0.510	4.81–4.25	5.40–5.27	0.559–0.582
B (40 μ)	9200–9490	4.51–4.98	1.055	1.75–2.84	4.57–4.43	0.336–0.367
C (60 μ)	9520–8750	6.89–8.03	2.060	15.5–17.3	5.36–5.95	0.273–0.267

Table 9. Results of the simulation (After [64])

Particle	Initial velocity (m/sec)	Initial Tempera-ture (K)	T_{max} (K)	Mass evapora-tion (%)	Residence time (m sec) (16 cm fall)	T_{exit} (K) (at 16 cm)	V_{max} (m/sec)
A (20)	5.0	300	2906 (6.1)	92.5	22.92	2084	16.5 (10.3)
B (40)	5.0	300	2902 (6.8)	21.6	24.55	2595	10.0 (12.7)
C (60)	5.0	300	2411 (14.3)	1.2	26.79	2400	7.75 (13.3)
D (80)	5.0	300	1759 (16.0)		27.95	1759	6.88 (13.7)

Results in brackets present the axial position (cm) where each maximum was computed

Fifty mm from the exit, particle and plasma temperature are the same. The evaporation process is over and the particle dimension is stabilized to about 45% of the initial radius.

All the theoretical results have been compared with measurements on 20–28 μm particles (dark circles) and 50–63 μm particles (open circles) Fig. 29a. If velocity measurements are in good agreement with the calculations, the measured temperatures are from 100 to 600 K lower than the predicted ones. However, it is to be noticed that the measurements represent mean values and that the particles in their totality do not follow the jet axis. The same calculations have been performed for a particle trajectory like that represented with shaded line on Fig. 25a. In this case a strong evaporation process occurs too, leading to a 50% particle radius diminution.

In Fig. 29b are shown the temperature distributions inside the particle for different axial positions. The characteristic positions are:

3 mm	Melting temperature at the external surface
8.45 mm	Boiling temperature at the external surface
10.7 mm	Total particle fusion
55.9 mm	Surface temperature = plasma temperature
90 mm	Plasma temperature is lower than melting temperature

Note that the core temperature is higher than the surface temperature.

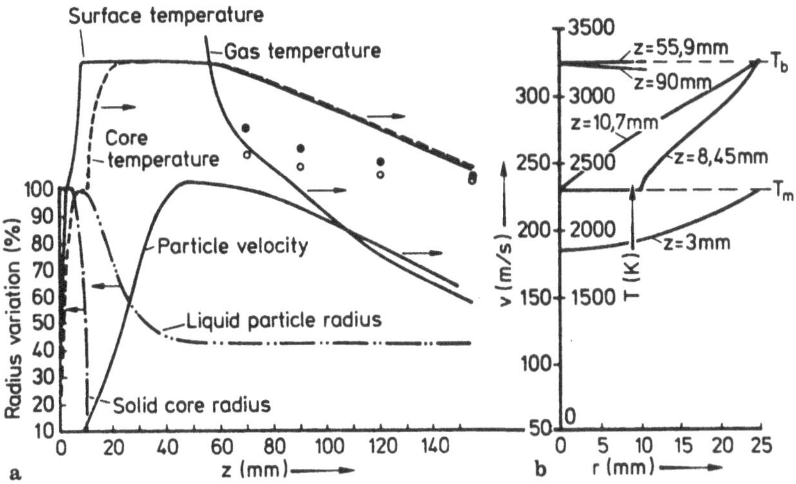

Fig. 29. Temperature distribution for Al_2O_3 particles in a DC argon-hydrogen plasma jet. Reprinted from [65]

c) *Some chemical reactions*

For the MoS_2 desulfurization process, Munz [71] has shown that the reaction

$$MoS_2 \rightarrow Mo + S_2$$

is heat transfer controlled and that MoS_2 decomposition starts appreciably in the solid state from 1773 K. In flight the moving particles go through the different phases summarized [71] in Table 10.

d) *Generator design influence* [66]

The thermal efficiency of plasma powder spraying is determined by the design of the plasma torch, which is strongly dependent of the material feeding means. In order to show the energy balance of plasma spraying processes, Rykalin [66] proposes the schemes presented in Fig. 30 and 31.

When material to be sprayed is supplied to the reactor in a wire form, use of the wire as the generator anode allows to convert about 10% of the electric power instead of 2–5% when the wire is placed in the free jet (Fig. 30a and b).

With powders, injection downstream of the arc leads to about 5% energy conversion. Injecting the powders through the arc close to the cathode leads to an energy conversion up to 30% (Fig. 31a and b).

Moreover, recently Vardelle [72] has shown the importance of the design of the nozzle of the plasma torch. One of the most important point to achieve the complete melting of the particle is the residence time. For example, with the same electrical power (30 kW) and the same volume flow rate either of $Ar-H_2$ or N_2-H_2 the temperature of the plasma along its axis at the nozzle exist is about the same but the velocity of 18 μm alumina particles injected the same way in the two plasma generators is 1.5 greater with the N_2-H_2 gas than with the $Ar-H_2$ gas. So the yield of melted particles with N_2H_2 is only 70% of that obtained with $Ar-H_2$. To

Table 10. Successive steps of the molybdenum disulfide desulfurization process [71]

Temperature (K)	1773	1773 to 1923	1923	1923 to 2892	2892	2892 to 4919	4919
Successive states or reactions	Solid state heating only of MoS_2(s)	1/Heating of MoS_2(s) 2/MoS_2(s) → → Mo(s) + S_2(v) 3/Heating of Mo(s)	1/Melting MOS_2(s) 2/MoS_2(s) → → Mo(s) + S_2(v) Mo stays at 1923 K	Solid—liquid state 1/MoS_2(l) → Mo(s) + S_2(v) 2/Heating of MoS_2(l) 3/Heating of Mo(s)	Melting of Mo(s)	Heating of Mo(l)	Vaporizing Mo(l) → Mo(v)

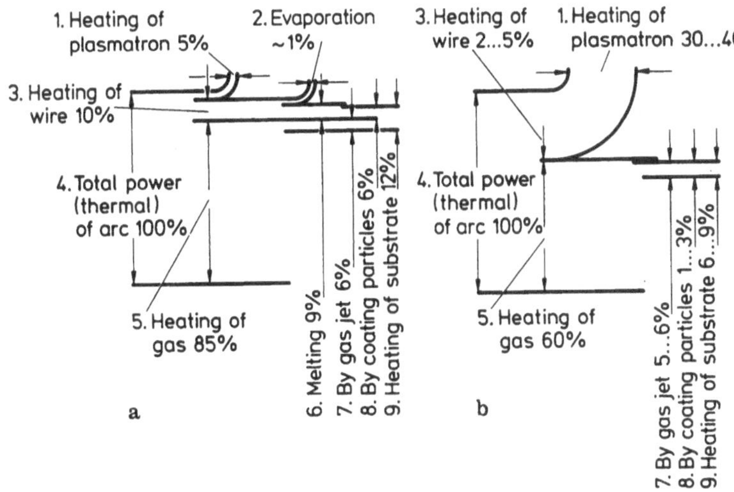

Fig. 30a, b. Energy balance of plasma spraying anode-wire **a** and neutral wire **b**: arc power — 10–12 kW; argon plasma; spraying distance 100 mm. Key: (1) heating of plasmatron, (2) evaporation >1% (3) heating of wire, (4) total thermal power of arc 100% (5) heating of gas, (6) melting, heating of substrate: (7) by gas jet, (8) by coating particles. Reprinted from [66]

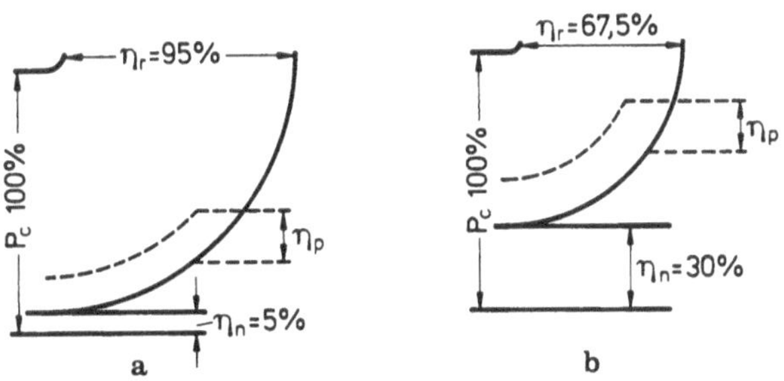

Fig. 31 a, b. Energy balance of plasma powder spraying: injecting the powder into plasma jet (at the edge of the nozzle), **a** to the arc column, **b** near the cathode: P_c — plasma energy: y_r/y_p and y_n — efficiencies of heating the plasma forming gas, sprayed substrate surface and particles (up to melting temperature). Reprinted from [66]

get about the same yield with the two mixtures, the diameter of the nozzle working with N_2—H_2 has to be greater in order to reduce the gas velocitv and, then to increase the particle residence time.

2.3.5 Chemical Kinetics

2.3.5.1 Introduction

When equilibrium laws cannot be applied, the chemical composition of a streaming medium may be obtained by the techniques of chemical kinetics.

Generally, one must take into account the elementary processes of collisions, physical and chemical kinetics, fluid dynamics and thermodynamic relations for the reactive flow. Chemical changes occuring in a streaming gas mixture (chemical reactions, ionization . . .) and energetic level changes in internal motions of molecule (or atom) depend on different characteristic relaxation time, t_p, to reach their equilibrium values. (t_p is, for example, the vibrational relaxation time).

On the other hand, we can define some mechanical times, t_m, as, for example, the residence time of a molecule in a reactive volume. With t_p and t_m we can form the t_p/t_m ratio, known as the Damköhler parameter [28], D_I:
if $D_I \ll 1$, the evolution time of the considered process is much lower than the typical associated mechanical time, the flow is assumed to be in equilibrium,
if $D_I \gg 1$ no chemical composition changes occur in the flow during the t_m time: the flow is "frozen",
if D_I is about 1, the flow is said to be "relaxed" and this relaxation is mainly the cause of chemical kinetics.

When a chemical system is submitted to certain physical conditions, its composition can be determined by thermodynamic means if each transformation is performed as a succession of equilibrium states. In fact some competitive reactions may occur and the evolution of the system is determined by the fastest reactions. If consecutive reactions take place, it is the slowest ones which govern the evolution and the system can stay in a metastable state during an undetermined time. Thus, as they do not take account of time, the thermodynamic laws are often used to provide the real composition of such a chemical system. But in the general case one have recourse to kinetic models which tend to predict the influence of physical conditions on the reaction rates.

2.3.5.2 *Kinetics Formulations*

In a stationary medium the variation by unit time of the ith specie concentration is given by [27]:

$$\frac{\partial C_i}{\partial t} = \nabla \cdot (D \nabla C_i) - P_i + S_i \tag{74}$$

where D is the diffusion coefficient, S_i is the source term and P_i is the losses term (chemical reactions). Disregarding the diffusion process because its contribution introduces many difficulties in the solution of (74), we have only to consider the contribution of chemical reactions in a homogeneous medium.

Chemical reactions occuring in a gaseous mixture involving I species A_i can be written in the following form:

$$\sum_{i=1}^{I} \nu'_{ji} A_i \xrightarrow{k_j} \sum_{i=1} \nu''_{ji} A_i \tag{75}$$

where k_j is the specific rate constant of the jth reaction, ν'_{ji} and ν''_{ji} are the stoechiometric coefficients. If we now consider an open system where the pressure remains constant as the temperature varies with time T(t), the variation in unit time of the ith concentration is given by:

$$\frac{dC_i}{dt} = w_i - C_i\left(\frac{kT}{P}\sum_{i=1}^{I} w_i + \frac{1}{T}\frac{dT}{dt}\right) \tag{76}$$

where w_i is the production-destruction rate of specie i, with:

$$w_i = \sum_{j=1}^{J} k_j(v''_{ji} - v'_{ji}) \prod_{l=1}^{I} C_l^{v'_{jl}} \tag{77}$$

Thus, the composition of such a chemical system may be obtained, at time t, by solving the set of non-linear differential equations (76). The solution is generally obtained by numerical integration, which must be highly stable ones, as the problems encountered are "locally exponential", giving rise to the phenomenon called „stiffness". Warner [73] gives a critical review of the different numerical methods employed to approach the solution of stiff differential equations.

However the major problem in chemical kinetics is due to a poor knowledge of the specific reaction rates k_j. These quantities are strongly temperature dependant.

The values of the rate constants k may be found in the literature for temperature lower than 3000 K. The values given by different authors may be very different arising, for example, because experimental measurements were not performed under the same conditions (crossed beams, shock waves, etc. . . . .).

From a theoretical point of view, the "reaction rate theory" attempts to determine the specific rates from the physical properties of the molecules or atoms involved

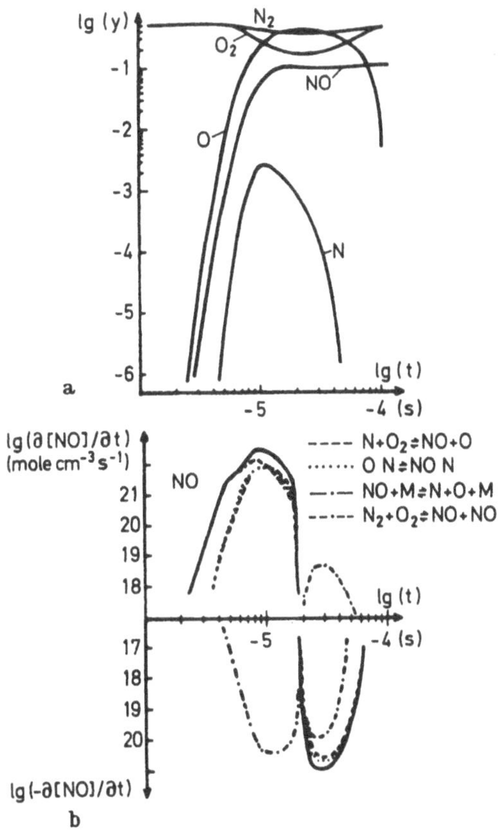

Fig. 32. Kinetic model for nitrogen oxides synthesis. Reprinted from [78]

in the reactions considered. Two methods have been developed in order to calculate the specific rates. They are the so called "collision theory" [74] and "activated complex theory" [74–76]. However this approach is highly complex and generally it does not yield k-values with a sufficient accuracy [77].

2.3.5.3 Examples of Kinetic Calculation

The calculations have been performed in the case of nitrogen oxides synthesized in a D.C. plasma jet [78].

The theoretical model attempts to evaluate the change in chemical composition of a gas mixture as a function of time at constant pressure. The temperature history is described by a parabolic law. The gas is heated up from 300 K to 5000 K in 10^{-5} sec with a 10^9 K/s initial heating rate. A similar parabolic law is used for quenching step. Temperature goes from 5000 K to 300 K in $9 \cdot 10^{-5}$ with initial rate about 10^8 K/s. The composition of the gaseous mixture is plotted as a function of time in Fig. 32a. Figure 32b describes the evolution of the NO production rate according to the different chemical reactions involved in the model. The specific reaction rates were those given by Baulch et al. [79].

This model shows the important influence of heating and quenching rates on the amount of final products. The heating rate controls the maximum NO concentration values as the quenching rate controls the freezing of the high temperature mixture.

2.4 Diagnostics

Under this heading we include, as usual, measurements of plasma gas temperature and velocity, of excited level populations and of electron density. In connection with ceramics, we will also include, in this part the measurement of temperature and velocity of condensed materials injected into the plasma gas. These measurements are generally rather complicated from both point of view of experiment and theory, but the results, even approximate, are of primary importance in understanding what happens during plasma treatment.

2.4.1 Plasma Diagnostics

2.4.1.1 Temperature and Population of Excited Levels

For plasma jets and RF plasma, the most often used diagnostic tool is spectroscopy because it does not disturb the plasma flow. The intensity of radiation emitted at a given wavelength is related to the temperature and excited level populations, but one must realize — what temperature means in this case.

Fundamental considerations

The degree of thermodynamic equilibrium has been extensively studied from a theoretical point of view by Griem [80] and Drawin [81]. It has also been studied experimentally and compared with theory, for hydrogen, helium, and argon [81–83] and recently for nitrogen [84]. One must bear in mind the following points:

for a given specie a Maxwellian distribution defines the translational temperature T_k, the Boltzman distribution of two excited levels i and j defines the excitation temperature T_{ex}, the mass action law defines the reactional temperature T_{re} of the concerned species, Planck's law defines the radiation temperature T_{rad}.

Complete thermodynamic equilibrium (CTE) is achieved if $T_{tr} = T_{exc} = T_{re} = T_{rad}$. CTE is not realized in laboratory plasmas wich are optically thin (Planck's law is not valid). But the mass action law, the Boltzmann distribution and Maxwell distribution may be obeyed by a unique local temperature such that $T_t = T_{exc} = T_e = T$, then one introduces the (complete) local thermodynamic equilibrium (LTE). Boltzmann and Saha's law are often obeyed only for highly excited levels, the plasma is then said to be in Local Partial Thermodynamic Equilibrium (LPTE).

The equilibrium conditions are generally stated from characteristic time and length for the process under consideration, compared to characteristic time and length of a plasma parameter. As an example for a cylindrical plasma of radius r, the characteristic length and time of the plasma temperature are:

$$L^T = T(r, t) \Big/ \frac{\partial T}{\partial r}(r, t)$$

$$t^T = L^T / \overline{v(r, t)}$$

where $\overline{v(r, t)}$ is the mean velocity. The following processes have to be studied:
collision times between different particles [12,85,86],
Maxwellisation times [87],
energy relaxations [88-92],
kinetics of reactions [84],
Boltzmann distribution relaxation [81],
diffusion lengths [81,93] and
equilibrium population of energy levels [81].

The characteristic parameters generally considered are the temperature and electron density which are of primary importance under equilibrium conditions [94]. For example, with a D.C. plasma generator, two different zones have been pointed out in a nitrogen plasma jet at atmospheric pressure [84]. The first zone is such that $10^{15} < n_e < 10^{17}$ cm^{-3} and 9000 K $< T <$ 15000 K, and the different criteria show that LTE is achieved. The second zone has a low electronic density $10^{12} < n_e < 10^{14}$ cm^{-3}, a quite low temperature 3000 K $< T <$ 7000 K, and equilibrium is not realized, due in part to diffusion. Nevertheless in both cases, the relaxation times of rotation-rotation, and rotation-translation exchanges are sufficiently low to consider the rotational and the translational temperatures to be equal.

b) *Spectroscopic measurement methods*

For spectroscopic diagnostics one must realize that the lines should have a high intensity compared to the continuum, they must be well separated, and at atmospheric pressure, an Abel's inversion [95] is generally needed.

The main diagnotics used are the following: The continuum emission due to free-free and free-bound radiation that is well suited for temperatures between 8000 K and 14000 K (the continuum intensity, proportional to $Ne^2/\sqrt{T}$, has been studied for nitrogen [96], argon [97], helium [97], hydrogen [98], and air [99]).

The ratio of two line intensities or the Boltzmann plot [95] gives an excitation temperature [84].

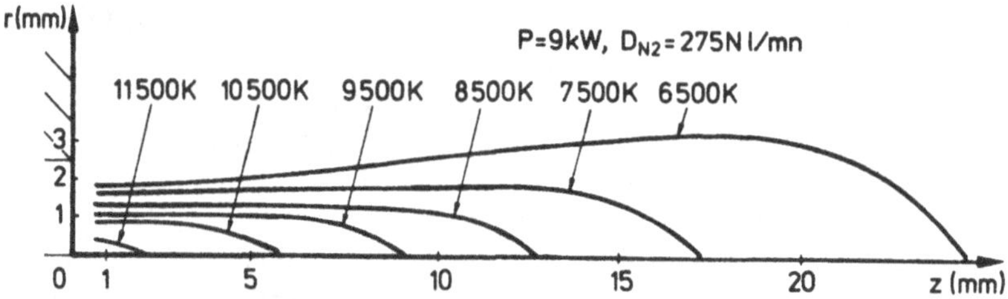

Fig. 33. Temperature contours in a DC nitrogen plasma jet. Reprinted from [84]

The atomic line profiles: when purely Stark, these profiles are well suited for the measurement of electron density (usually one uses Stark width of hydrogen or helium lines with the gases introduced in small quantity (less than 0.5%) in the plasma gas) [80], when purely Doppler, these profiles give the neutral temperature; when the line profile is due to contributions from both, Stark and Doppler broadening must be separated, and one needs a high resolution apparatus. The most often studied profiles are those of argon [100], helium [101] and nitrogen [84].
The absolute intensity of lines can be used to measure the upper level population of the transition considered when the excitation temperature is known. The molecular band spectra are used to measure the rotational temperature (equal to the translational one) when the lines are sufficiently separated, but even when the lines are not separated, comparison between the experimental profile and a set of calculated profiles as a function of temperature give rather good information especially below 6000 K [95, 84, 102].

Figure 33 shows, for example [84], a temperature chart of a nitrogen plasma jet; these measurements have been performed spectroscopically with an automatic data

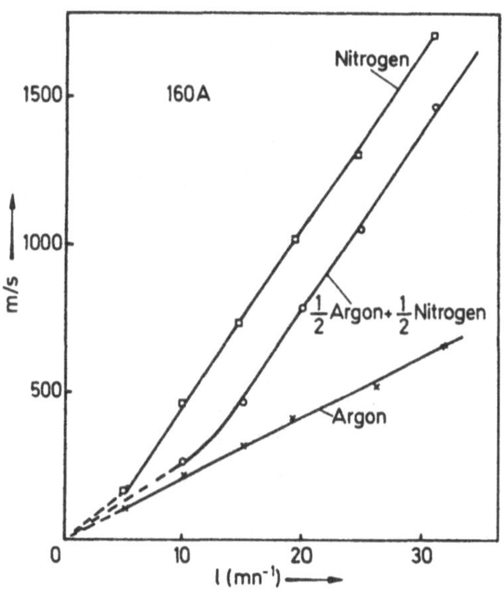

Fig. 34. Typical measured axial velocity of DC plasma jet versus gas flow rate for different gases (argon, nitrogen and their mixture). Reprinted from [106]

acquisition system connected to a computer, which reduces the measurement time to a few hours.

2.4.1.2 Plasma Gas Velocity Measurements

Under the assumption of LTE a rather simple method consists in measuring the dynamic pressure of the gas with a probe. But, in that case, the probe disturbs the plasma flow, even if it is precisely profiled [103,104]. Another technique consists of the observation of a small electrical perturbation superimposed on the discharge [105,106]. The use of a pulsed laser now offers new opportunities for this type of measurement [107]. Figure 34 shows the velocity of an argon and nitrogen plasma jet at the nozzle exit as a function of the gas flow rate. A third technique uses very small particles injected into the plasma. In HF plasma the particles velocity, supposed to be the same as the gas flow, is measured by laser anemometry [107,108]. However for DC plasma arcs the injection of small particles directly into the arc raises complex technical problems [109]. Finally for very high velocities (higher than 2000 m/s) it is possible to utilize the Doppler shift of the lines [110].

2.4.2 Condensed Particles Injected into the Plasma

2.4.2.1 Velocity of Condensed Particles

There are three kinds of measurements methods. The first kind is mechanical methods in which the flow of particles is momentarily stopped with a barrier, and then one observes the rate of the flow of particles downstream. With this method

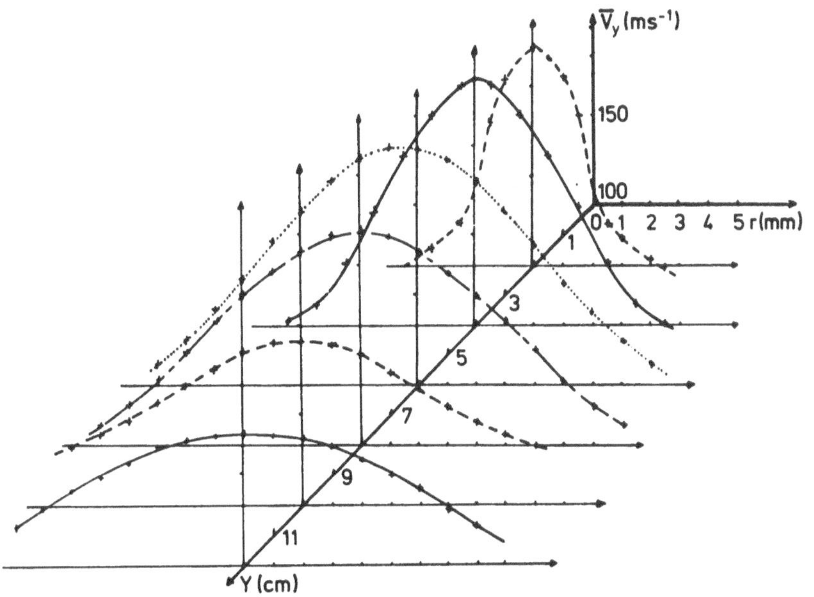

Fig. 35. Radial alumina particle (φ 18.5 μm) velocity distribution along a DC nitrogen-hydrogen plasma jet. Reprinted from [109]

a mean velocity is obtained in a plane orthogonal to the flow direction. With two holes, in two discs it is possible to obtain more refined measurements. With this type of technique it is possible to measure velocities between 50 and 200 m/s [111] but the flow is strongly perturbed. The second kind is optical methods which have been until now mostly used. An image of the jet is photographically recorded either with a rotating mirror [65,112] or an ultra rapid camera [113]. Velocities up to 750 m/s have been measured by this technique. The third kind is called opto-electronic methods in which one uses optical detection of the particles connected to an electronic system for data treatment. This is for example the case of the "plasmascope" developed by Gold [114]; the "plasmascope" makes an analysis in space and time of the luminous perturbation of particles flowing in the plasma. But now a more sophisticated method is available in Doppler laser anemometry [115,116]. Many techniques of this type have been developed and give fairly good results. The velocity of particles can be measured at any point in the plasma [117–122]. For example, Fig. 35 shows [109] the axial and radial velocity of alumina particles injected in a nitrogen-hydrogen plasma.

2.4.2.2 *Temperature of Condensed Particles*

The analysis of the radiation emitted from the surface of the particle gives the radiation temperature. This method has been used by Bonet for low velocity particles (< 30 m/s); a monochromatic photograph of the moving particles is recorded through a pyrometer which gives the reference radiation [123]. Lesinski [124] has developed a similar method by using a photomultiplier in place of camera; the reference radiation is given by a hole travelling very rapidly in front of a reference W ribbon lamp. This method has been refined by Vardelle [109], who has made a statistical treatment of the received signals. The main difficulties with this system arise from the need to know exactly the particle diameters, and their emissivities.

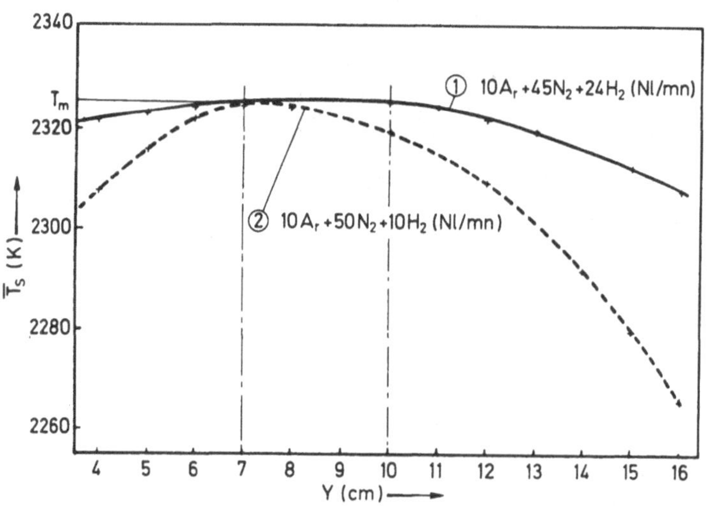

Fig. 36. Alumina particle surface temperature distribution along a nitrogen-hydrogen DC plasma jet. Reprinted from [109]

Figure 36 shows the temperature of alumina particles along the axis of a nitrogen hydrogen plasma jet.

3 Plasma Generation

For any type of process which can be developed on an industrial scale, one must be able to plan all sorts of scaling on the economical and technical points of view. For this reason plasma torch or plasma reactor modeling has become an important study. For example Lonza Corp [125] has proposed a study of the relative costs of different plasma torches that we have actualized (Table 11). In this paper we present an outline of the main plasma devices used in the ceramics field.

3.1 HF Torch

A plasma column can be stabilized in a non conducting tube (silica, boron nitrides, etc. . . .) by RF inductive or capacitive coupling (between 1 and 20 MHz). The main advantage of this technique arises from the fact that the plasma column is not in contact with electrodes, allowing, in particular, the use of aggressive gases such as oxygen, chlorine [126] or UF_6 [127]. With a life time of 2300 hours an HF plasma device can work with a power up to 1 MW (Ionarc Tafa torches), but in the configuration presently employed it is not possible to transfer more than 40–60% of the electrical energy to the plasma. However, the efficiency can be increased up to 70 to 80% [128] by operating the generator under overload conditions or using special tubes with magnetic focusing.

3.1.1 Torches with Inductive Coupling

This technique is most widely used at pressures between a few Torrs and atmospheric pressure. For example, Fig. 37 shows a 30 kW torch used by Hamblyn [129] for the reduction of volatile metal chlorides. The oscillating magnetic field is produced by a coil, external to the water cooled silica glass envelope and supplied with a HF current from a vacuum tube generator. Practically any gas (often mixed with argon) may be used in such a device, the technical problems (overheating of the tube) arising with pure hydrogen are solved by using special cooling with copper tubes. Generators with power output of 200–300 kW are currently used [128] for the preparation of titania pigment (oxygen plasma with $TiCl_4$) and, as already mentioned, units with power up to 1 MW are used. The minimum power necessary for self-sustained induction discharge is determined by the nature of the gas, its pressure and the frequency of electromagnetic field [128]. As the frequency is reduced from the MHz range to the hundreds of kHz range, the minimum power increases from less than 10 kW to hundreds of kW and then rises hyperbolically with further reduction in frequency. To reduce this minimum power one has to increase electrical conductivity by decreasing the pressure or adding ionizing impurities (K, Cs, etc. . . .). Although phenomena associated with the HF plasma are now well known in principle, modelling [130] is very difficult to carry out because the high temperature gradients (see Fig. 23a) result in large variations of the transport properties and torches

Table 11. Composition of Plasma Generators [125]

			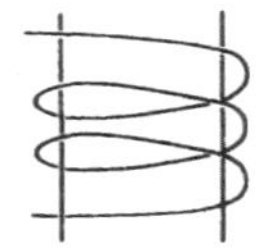	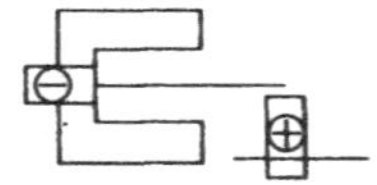
Type	DC-ARC	3-Phase AC-ARC	Induction ARC	Liquid Stabilized DC-ARC
Availability of industrial units	150–200 kW (C-ARC: MW)	500–1000 kW (C-ARC: MW)	100–1000 kW	300–500 kW
Investment cost (Torch and controls) (200 kW Unit)	$ 900/kW	$ 1000/kW	450 kHz $ 2000–2500/kW 10 kHz $ 900–1200/kW	$ 1000/kW
Operating (stabilizing) medium	Noble gases, N_2 H_2 (hydrocarbons)	Noble gases, N_2 H_2 (hydrocarbons)	any gas or vapour	any insulating liquid (water, alcohols, hydrocarbons etc)
Chemical nature of plasma	neutral, reducing	neutral, reducing	neutral, reducing, oxidising, reactive	neutral, reducing, oxidising (reactive)
Torch efficiency	60–80%	80% +	50% +	85–90%
Power source efficiency	90–95%	95%	30–80%	90–95%

The prices given by Schnell in 1974 have been actualized on the basis of the actual cost in France.

* C for Carbon

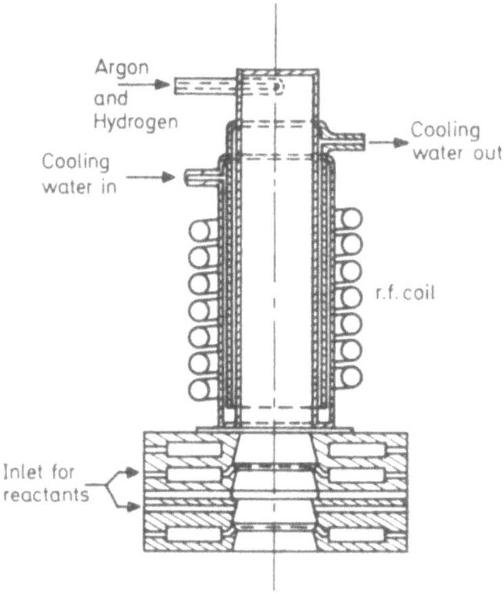

Fig. 37. A 30 kW radio-frequency induced plasma reactor. Reprinted from [129]

must be essentially designed in a qualitative manner. Finally, it must be mentioned that the gas velocity in a HF torch is much slower (several tens of m/s) than in arc torches [61] (hundreds of m/s). Furthermore, the diameter of the reactor is important (20–200 mm) and the energy density is much lower than in arc plasma generators and lower than in arc plasma furnaces.

3.1.2 Generators with Capacitive Coupling

Much less common is capacitive coupling with the electrodes external to the insulating tube leading to the formation of a phase shift between the electrodes and discharge current thus reducing the efficiency of the discharge. However the minimum power necessary for a self sustained discharge is lower than that for magnetic coupling. Rykalin [128] mentions, for example, that in the range 10–20 MHz, this minimum power is equal to 0.2 kW for air and 1.0 kW for hydrogen operation. Intensive work on this type of discharge is in progress at Baikov Metallurgical Institute of Moscow.

3.2 DC or AC Plasma Torches

3.2.1 General Considerations

The first design of a plasma torch was given in 1957 by Gage [131] who used a direct current arc struck between a cathode rod and a nozzle anode. Forced gas flow extended the arc in the anode nozzle which was strongly cooled. A thermal arc pinch effect was produced by the joint action of the cold wall arc channel and the cold gas sheet around a very high temperature conducting core (the arc column).

Various torch configurations are possible depending upon the arc stabilization mode: tangential gas input in the arc channel (Fig. 38a), axial gas input along the

a
b
c
d
e

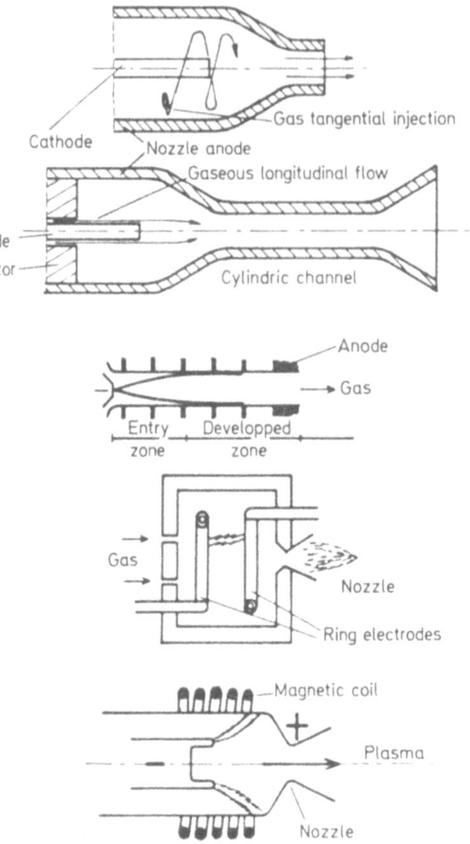

Fig. 38a—e. Principles of DC plasma torch stabilization
a arc plasma torch with stabilization by eddy gaseous flow
b arc plasma torch with stabilization by longitudinal gaseous flow
c wall stabilized arc
d autostabilized arc
e external stabilization

cathode (Fig. 38b), segmented anode arc (Fig. 38c), magnetic rotation of the arc root. The magnetic field can be self induced (by an arc current greater than 8000 A), or externally generated (Fig. 38d and 38e).

The anode nozzle is generally made of copper, molybdenum or carbon. Thermal losses can be calculated from the formulas given in 2.3.2.4. Depending on the nature of the gas and on the working parameters anode losses can range between 80% (laminar flow) and 10% of the energy input in the arc. Some empirical formulas giving the anode losses have been proposed [132–135]: the anode losses are proportional to the current intensity and are a function of the arc voltage to the power of 0.2 to 0.4. For a given power one can thus reduce the anode losses by using a large arc voltage and consequently, with a long arc. As mentioned in Ref. [136] the arc can be divided into three zones:

a) the entrance zone characterized by the formation of an hydrodynamic boundary layer along the arc channel;
b) a middle zone where interactions between the hydrodynamic boundary layer and thermal boundary layer occur;
c) and a turbulent flow zone.

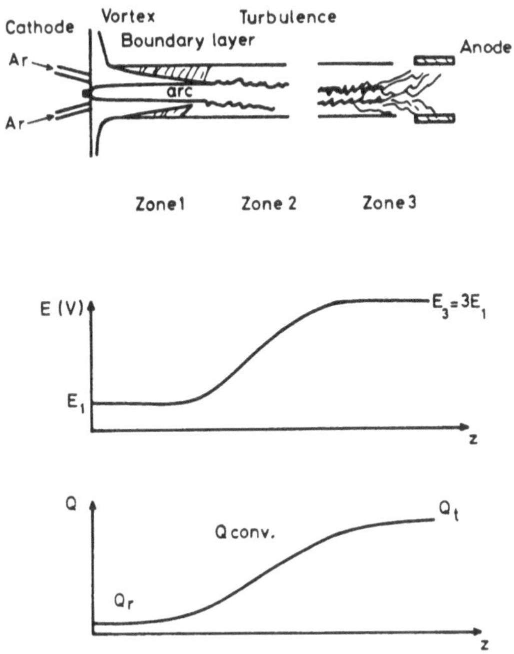

Fig. 39. Development of an extended arc, variation of the electric field and heat loss in the anode

These three zones are characterized by different properties as shown in Fig. 39. Contrary to the situation in 2nd and 3rd zone there is no radial pulsations of the arc in the first zone; in the 3rd zone the arc becomes filamentary and divides into secondary arcs. In order to reduce the losses to the cold walls, the arc length is frequently limited in the first two zones; however, to realize this condition the nozzle design must be carefully made. With good stabilization, and in the calculated power range, anode erosion occurs mainly at the arc root. In the region of anode fall the life time of striking sites [137)] ranges between 10^{-4}–10^{-5} s, with a thermal depth of 5 μm (this is the reason why the arc root must be continuously moved). For copper anodes, the average electrode erosion ranges between 10^{-6}–10^{-7} g/Clb and the anode life time is consequently 300 h to 400 h. One can find in specialized literature a great number of proposed arc column models which can generally be applied to the case of plasma torches [138–142)]. It has been shown that it is possible to design the dimensions of an anode nozzle with quite simple calculations as a function of desired working parameters [143)] such as: mean mass temperature, arc current and nature of the plasma gas.

The cathodes can be of the hot type such as tungsten, carbon or molybdenum, cathodes which obviously must be used in a non oxidizing atmosphere. Under certain conditions of oxidizing atmosphere one can use zirconium or hafnium cathodes. The so-called "cold cathodes" are generally made of copper [137, 144–146)]. Heat losses at the cathode are generally quite low (less than 15% of power input to the arc). For a tungsten cathode tip, erosion with current is of 10^{-9}–10^{-10} g/Clb so that its life time is approximately 150 hours [147)]. Hot cathodes (thermoemissive cathode) are generally not used with A.C. torches which are commonly based on copper electrodes.

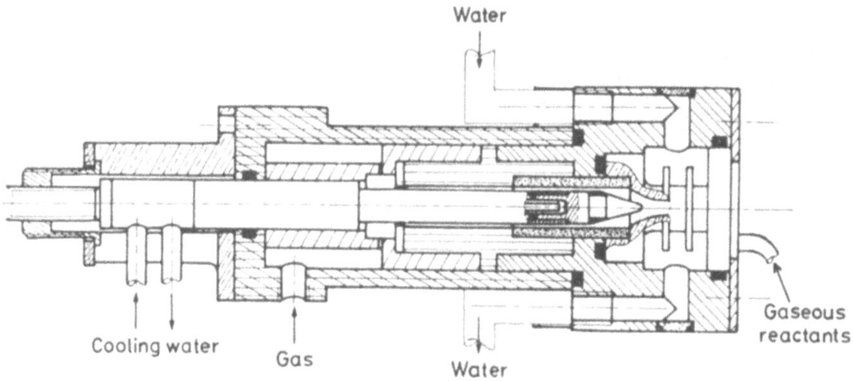

Fig. 40. A DC plasma torch. Reprinted from [84]

3.2.2 DC Torch

The most often used DC torch design is the one shown in Fig. 40. This type of torch which is used for plasma spraying is available as a general industrial tool for powers up to 100 kW.

For high power torches it is necessary to enhance the cooling of the electrodes. Figure 41a shows a torch with a variable nozzle diameter [147] which presents an increasing voltage — current characteristic. Figure 41b shows a torch with gas injection points in the arc channel; this type of torch works with power up to 500 kW at the Institute of Thermophysics of Novossibirsk [148] and it is planned to extend this to 50 MW. The following empirical formula was used to calculate the yield of the torch.

$$\frac{1-\eta}{\eta} = 5.85 \times 10^{-5} \left(\frac{I^2}{Gd}\right)^{0.265} (p \cdot d)^{0.3} \left(\frac{G}{d}\right)^{-0.265} \left(\frac{1}{d}\right)^{0.5}$$

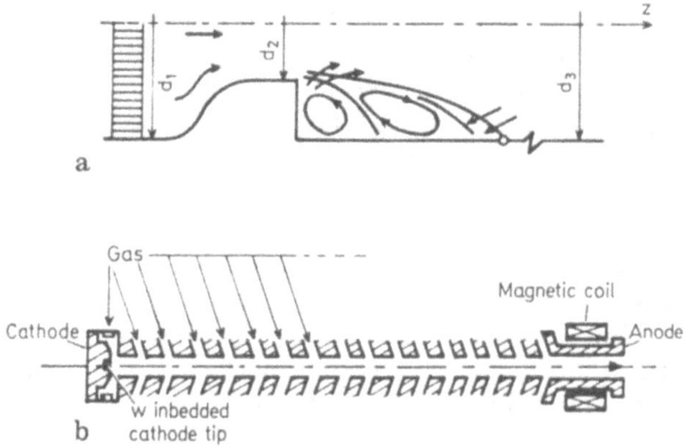

Fig. 41. DC plasma torch configurations: a nozzle with a change of diameter, b constricted arc gas heater with multi-injection. Reprinted from [147]

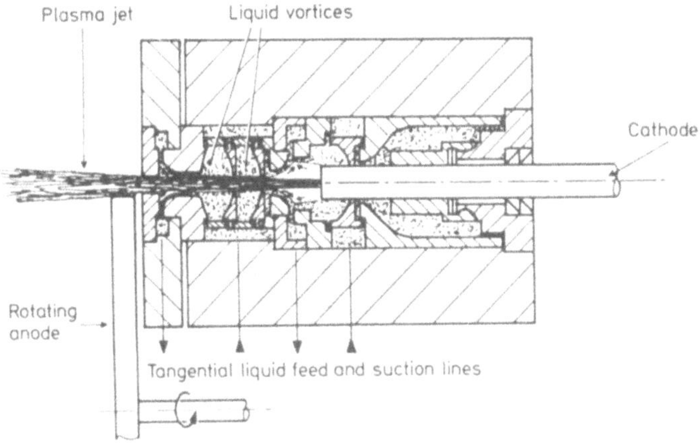

Fig. 42. Liquid stabilized plasma torch. Reprinted from [125]

In this formula η is the thermal efficiency of the torch, I the arc current, G the mass flow rate of the gas, d the diameter and l the length of the anode nozzle, p the gas pressure in the arc chamber. All these quantities must be expressed in SI units, as I^2/Gd in $A^2\ s\ kg^{-1}\ m^{-1}$, G/d in kg/ms and pd in N/m. This formula is valid under the following conditions:

$$5\ mm < d < 80\ mm; \quad 1 < p < 50\ atm; \quad 5 < G < 5000\ g/s;$$
$$5\ kW < P < 50\ MW; \quad I < 6000\ A$$

The torch using coaxial copper tube electrodes should also be mentioned in which the arc rotates under the influence of an external magnetic field and tangential gas injection.

Lonza Corp [125] has developed a 250 kW torch stabilized by an alcohol flow

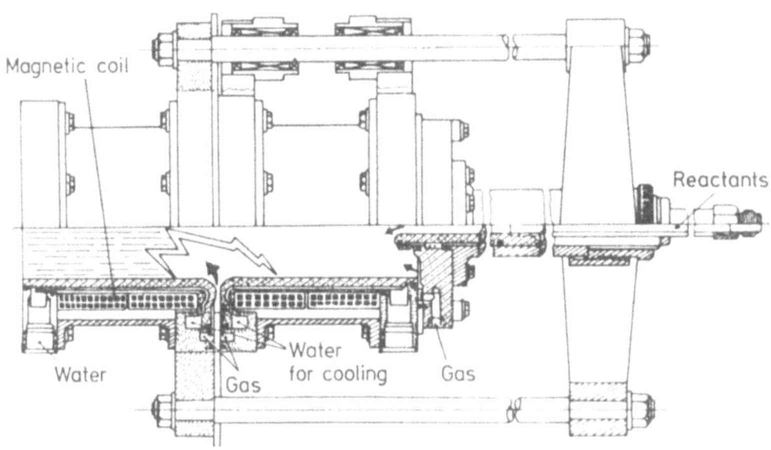

Fig. 43. A 3.5 MW single phase alternating current plasma torch. Reprinted from [149]

with a consumable carbon cathode and an external anode which is a copper-cooled rotating disk (Fig. 42).

3.2.3 AC Torch

In most cases the electrodes are made of copper, the arc strikes between two tubes as in the design proposed by Fey [149)] with power up to 1 MW (Fig. 43). The arc rotates under the influence of a magnetic field and of a tangential gas inlet. In order to avoid arc extinction a high frequency voltage is superimposed on the arc current [150,151)].

3.3 DC or AC Plasma Furnaces

In most cases these furnaces are used for the treatment of condensed materials. The main problem is generally the transfer of a sufficient quantity of energy from the plasma to the condensed material to achieve a desired treatment. We have seen in part 2.3. how a typical heat transfer calculation is performed, but the results have to be realized by experiment. The design of the furnaces described hereafter have been studied in this way.

Many bibliographic studies have been devoted to the different types of plasma furnaçes [152-155,128)]. In particular the review of Bonet [156)] provides a classification of plasma furnaces in which the reactants are fed as dispersed materials (this is generally the case of plasma ceramic treatment). Bonet assumes that, in a very simplified way (see 2.3.), the energy transfered to spherical particles of constant diameter d is expressed as follows:

$$E = \frac{6}{\varrho d} \int_{0}^{t_s} (h_T(T - T_p) + \sigma\varepsilon (T_0^4 - T_p^4))\, dt$$

where h_T is the convective exchange factor, T the plasma temperature, T_p the surface temperature of the particle, T_0 the wall furnace temperature, ε the emissivity of particles (the wall being supposed to be a black body) σ the Stephan constant, ϱ the mass density of the particles and t_s the residence time of the particle in the furnace.

Obviously the most effective thermal treatment is realized with a high E, which implies an increase of the residence time t_s of particle in the plasma, of the convective exchange factor h_T, of the wall temperature (to decrease radiation losses) and also a decrease of the particle diameter]

However, some of these considerations are not compatible. For example, h_T can be increased by increasing plasma velocity, but simultaneously the residence time will be decreased. The particle diameter cannot be decreased under 10 μm, firstly because it is difficult to transfer small particles in the plasma, and secondly because the milling cost would be prohibitive. Bonet [156)] has summarized in Table 12 some characteristic furnace parameters allowing a crude classification in three types A, B, C. We suggest the addition of a fourth D type with about the same residence time as C type but for which the charge is used as an electrode (falling film furnace,

Table 12. Classification of the plasma furnace (ref. [156]) as a function of the particle residence time (t_s) and diameter (d), of the plasma temperature T_p and of the plasma furnace wall temperature T_0

t_s (sec)	d (10^{-6} m)	T_p (K)	T_o (K)	Furnace type
10^{-3}	100	10^4	low	A
$10^{-2} - 1$	100	$5 \cdot 10^3$	low	B
			$2 \cdot 10^3$	B'
$1 - 10^2$	∞	$5 \cdot 10^3$	$2 \cdot 10^3$	C

usually DC anode), thus receiving additional heat from the arc which is partly lost in other configurations.

We shall present some of the main types of reactors that can be used for ceramic treatment. One can roughly consider a mean gas temperature around 6000 K and a velocity of a few m/s.

3.3.1 A Type Reactors

Most of the A type reactors are based on magnetohydrodynamic pumping near electrodes. This effect first observed by Maecker [157] is used to drag the particles very rapidly into the arc zone. Some calculations by Pateyron [158] using Bhattacharya's equations [159] show that for an H_2 plasma, atmospheric pressure, the local pressure reduction from the Maecker effect is about 2/1000 atm at 500 A. The main furnaces of this type are:
the expanded reactor of Ionarc [160] composed of a shielded cathode and three consumable carbon anodes (Fig. 44). The maximum power level of this reactor is 300 kW;

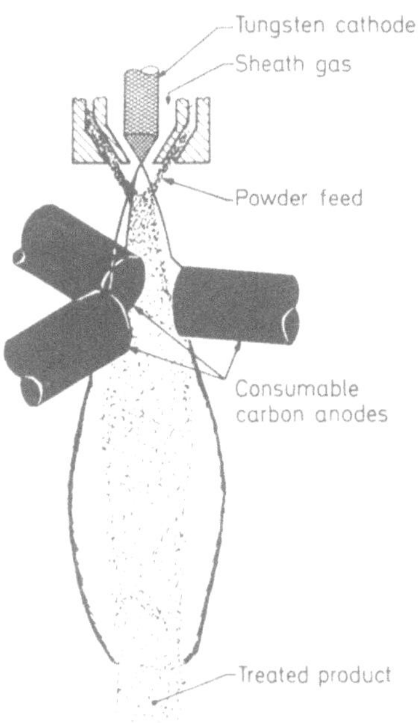

Fig. 44. Ion-arc heater for particles treatment. Reprinted from [160]

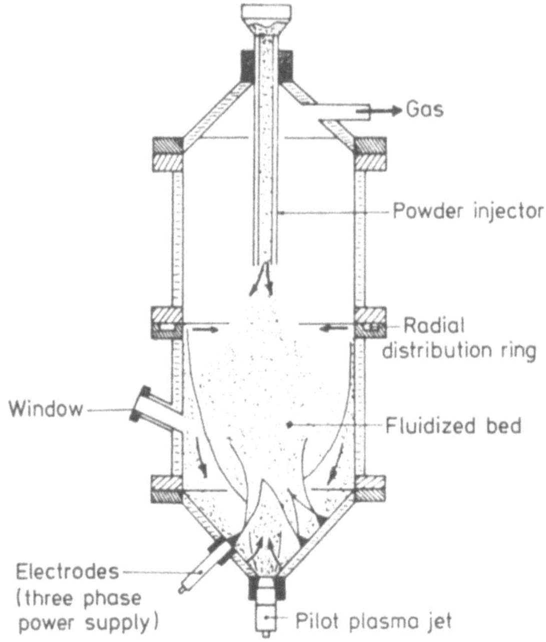

Fig. 45. Fluidized bed plasma furnace. Reprinted from [164]

the 50 kW reactor of Sheer [161] where the carbon anodes are replaced by three transpiration anodes;
the 100 kW reactor of the National Physical Laboratory in UK [162] where the anodes are three DC plasma jets, enabling higher power to be attained;
the Noranda [163] furnace where the arc strikes between a shielded cathode and an anode placed in a crucible where, for example, mölten molybdenum is formed.

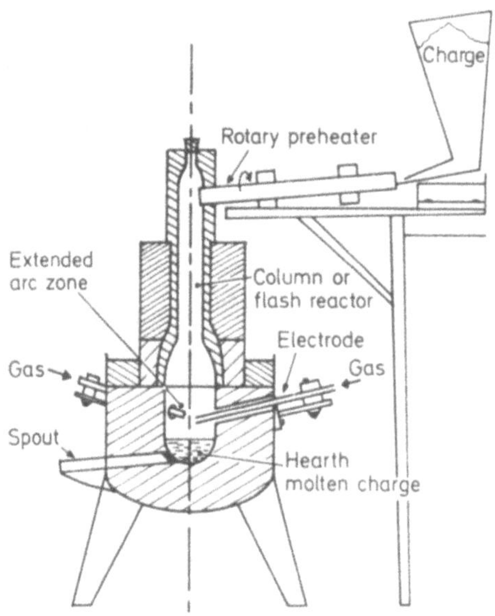

Fig. 46. Extended arc flash reactor. Reprinted from [165]

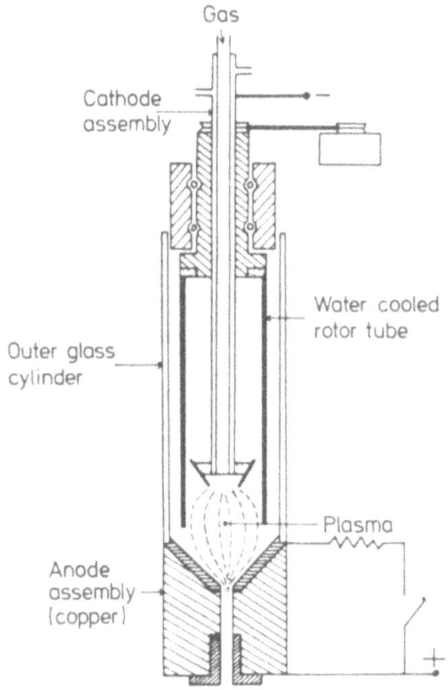

Fig. 47. A 30 kW rotating wall direct current plasma furnace. Reprinted from [166]

3.3.2 B Types Reactors

These furnaces are designed to ensure residence times of reactants of the order of 10^{-2}–1 s either with low (B) or high (B′) temperature walls. For this type of reactor one can distinguish two groups: the first group consists of sedimentation furnaces in which the solid reactants are fed either in co-current or in counter-current stream. We can mention Bonet's [164] furnace (B) working with a three phase alternating current in a cold-walled chamber. This furnace (50 kW) is mainly used for spheroïdisation with a counter-current feed (Fig. 45). The Segworth's [165] furnace is a

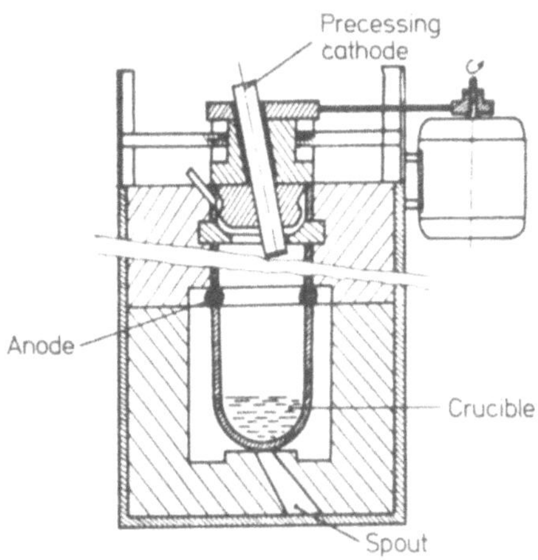

Fig. 48. A 200 kW rotating cathode plasma furnace. Reprinted from [167]

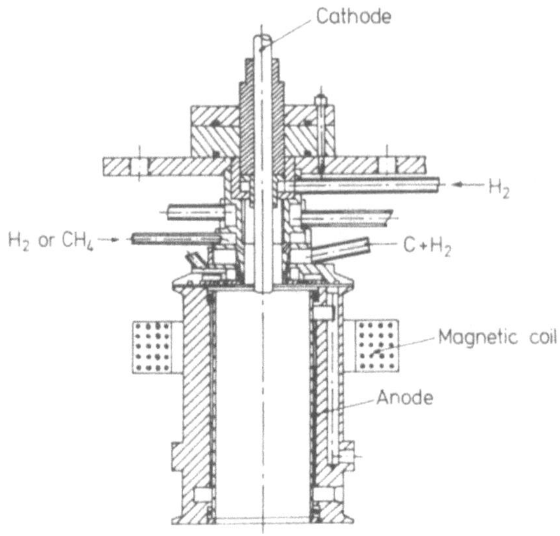

Fig. 49. The AVCO plasma furnace. Reprinted from [169]

30 kW hot-walled reactor (B′) using three consumable carbon electrodes. It has been used for chromite ore reduction (Fig. 46).

The second group consists of the so called expanded plasma furnaces. In this type of reactor the contact between the particles and the plasma is improved by expanding the plasma by various means:

(i) in the 30 kW DC plasma reactor of Whyman [166] expansion of the plasma flame is ensured by rotating the water-cooled walls (B) (Fig. 47).

(ii) in the 200 kW expanding precessive plasma reactor of Tylko [167] the expansion is realized by a precessive movement of the cathode in front of a toroidal anode (B′) (Fig. 48). It is worth noting that this furnace is supposed [168] to be scaled up to 1.4 MW by Foster Wheeler.

(iii) in the 100 kW plasma reactor AVCO [169] for acetylene synthesis (B) from coal (Fig. 49) expansion of the arc and improved contact of plasma and particles are realized with a magnetic field and a tangential gas feed.

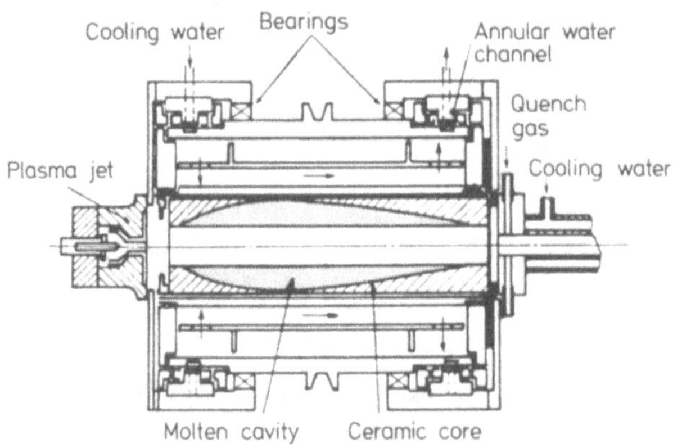

Fig. 50. A 30 kW horizontally rotating plasma furnace. Reprinted from [170]

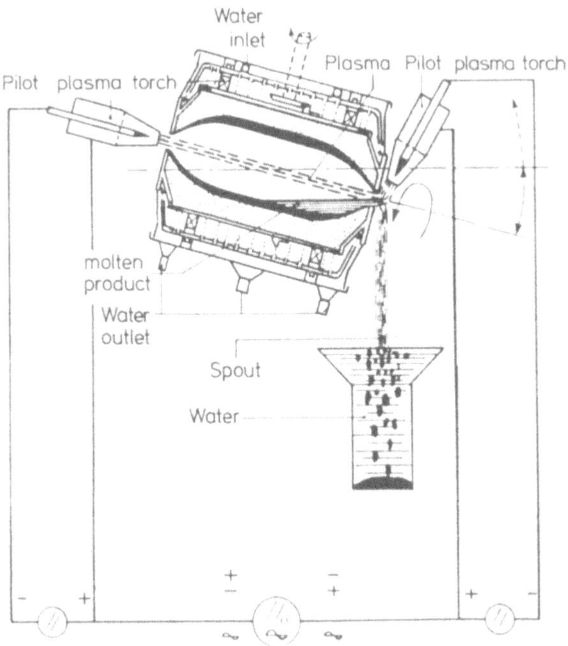

Fig. 51. A 100 kW sloping rotating plasma furnace. Reprinted from [173)]

3.3.3 C Type Reactors

This type of reactors has been designed to ensure a very long residence time for reactants which have to be melted to form a liquid film on the walls. A great number of reactors of this type have been proposed and, as in the preceeding type, they can be divided into two groups.

The first group consists of the rotating plasma furnaces:
The horizontal axis rotating furnaces were developed successively by Grosse [170)]

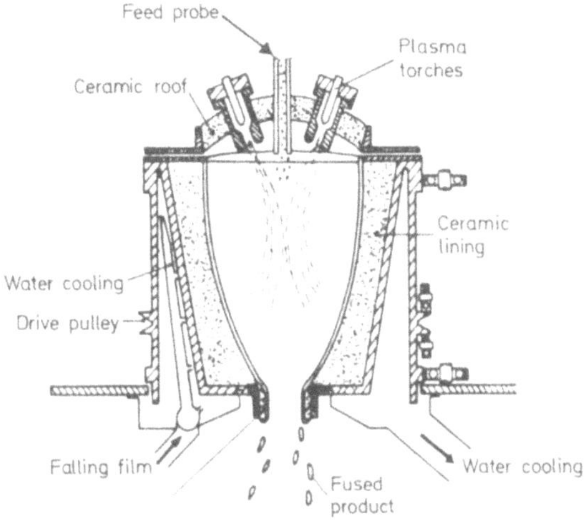

Fig. 52. A 60 kW vertically rotating plasma furnace. Reprinted from [174)]

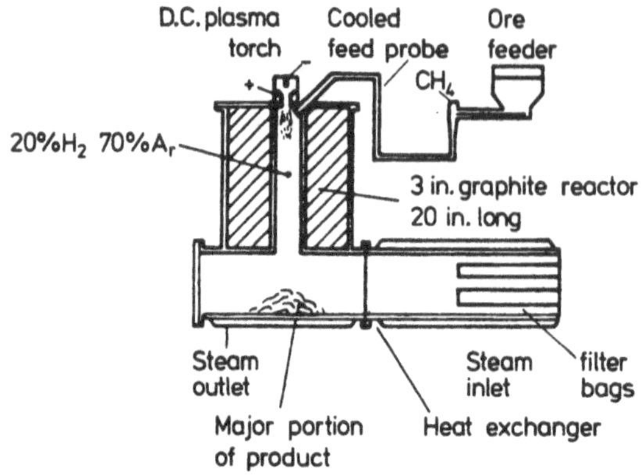

Fig. 53. A 50 kW American Cyanamid falling-film plasma reactor. Reprinted from [175]

in USA, by Foex [171] in France and by Sayce [172] in UK. The plasma flows in a cylindrical cavity made with the treated material, the outside wall being water cooled. When melted the material is held by centrifugal forces on the inside walls of the furnace. The plasma is generated by a single DC plasma torch, for example the 30 kW furnace of Grosse [170] (Fig. 50) or the 250 kW Lonza [125] furnace.
The 100 kW tilted rotating drum furnace developed by Yerouchalmi [173] is heated by a transferred plasma column generated between two DC plasma torches along the furnace axis (Fig. 51).
The 60 kW vertical rotating drum furnace developed by Sayce [174] is heated with two of three DC plasma jets flowing downwards to the liquid walls Fig. 52). In this arrangment a problem arises from radiative heating of the vault by the parabolic walls.

The second group consists of the so called falling film furnaces (Fig. 53) first patented by Chase [175]. By tangential feeding, the particles are centrifuged on to the reactor walls where they melt and form a slowly falling film of molten material. This arrangment ensures very long residence times (up to a few seconds).

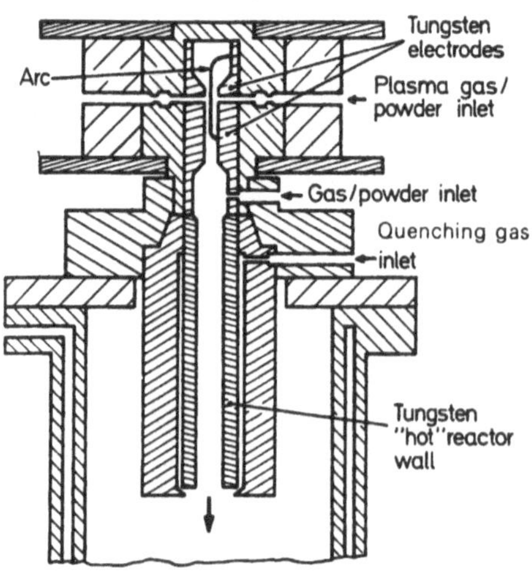

Fig. 54. Plasma reactor for silica vaporization. MHD Research inc. Reprinted from [176]

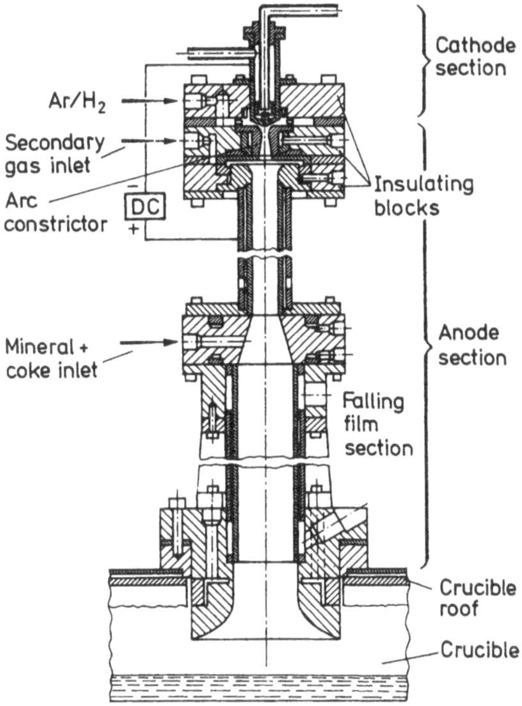

Fig. 55. Bethlehem Steel's MW falling film furnace. Reprinted from [177)]

3.3.4 D Type Reactors

In this type of reactor the charge is used as an electrode. It must then be pointed out that the melt is heated by the plasma as in other configurations but also heated by Joule effect and electron condensation (anode) (see Sect. 2.3.2.4.) thus enhancing the efficiency of the heat transfer to the reactants and reducing the heat losses to the electrodes.

One of the first furnaces of this type has been designed for MHD [176)] Research. This device is, in fact, a DC plasma torch with hollow cylindrical tungsten electrodes operating vertically (Fig. 54). Feeding of particles is realized in a gas vortex into the inner wall of the lower electrode (which is usually the cathode). The particles melt and form a liquid falling film.

MacRae et al. [177)] have built an improved 1 MW version of the Chase furnace (Fig. 55). Particles are fed tangentially to the wall where they melt and form a liquid falling film on the anode.

Though they are essentially used for metallurgical treatment we must mention the remelting furnaces such as the 60 kW Linde [178)] furnace which uses a DC plasma torch in place of the conventional graphite electrode (it is now planned to scale this up to 10 MW [179)]), and the furnaces which operate with ingot feeds heated with two or three plasma jets [180–182)]. This kind of reactor may be used for ceramic treatments if the charge is primarily heated by a plasma jet and used as an electrode when melted. They have applications in the melting of ceramic oxides such as ZrO_2,

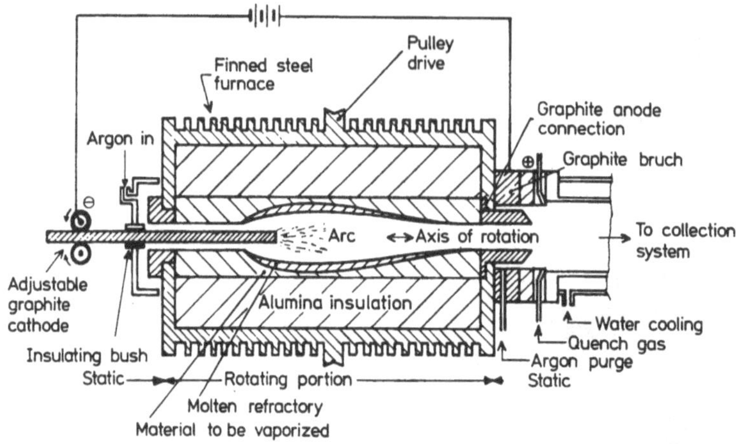

Fig. 56. Air-cooled centrifugal furnace heated by transfered DC arc. Reprinted from [183]

Al_2O_3, etc. . . . and for the manufacture of cast refractory blocks, abrasive and refractory grains and refractory fiber products.

The horizontal rotating drum furnace (Fig. 56) can be heated using consumable carbon electrodes [183]. The arc is initiated between a carbon exit nozzle and a cathode which can be withdrawn through the furnace as soon as the contents are sufficiently melted to be conductive and maintain the arc.

We finally have to mention the earliest devices developed by Sheer [184] on the basis of the well known high intensity arc where the material which has to be melted is compacted with carbon to form a consumable anode. However, as a consequence of excellent thermal heat transfer the use of this arrangement is limited by the thermal shock resistance of the electrodes.

4 Plasma Spraying, Spheroidisation, Vaporisation, and Condensation

4.1 Plasma Sprayed Coatings

The formation of protective coatings by spraying a stream of molten metal or ceramic particles was first developed using combustion flames into which the spray material was fed as a powder, wire or rod. In the 1960's commercial plasma spraying equipment became available in which a D.C. plasma jet was used to melt a powder feed and project the droplets at high velocity against the material to be coated. The major advantage over the flame spraying process is the higher particle velocity obtainable but the high temperatures achieved in the plasma jet also make it possible to melt even the most refractory material to produce high quality coatings. Plasma spraying is therefore particularly suitable for the formation of ceramic coatings for wear, thermal and corrosion protection. The plasma generator used are the DC plasma generators, described in Sect. 3.2.2. a, usually with power less than 90 kW, the particles being injected near the nozzle exit.

Plasma spraying has developed, to a large extent, by empirical means with relatively little scientific understanding of the mechanisms involved in coating formation and of the factors controlling the structure and properties of the coating. This is because the range of temperatures and the time scale of the various events is generally outside the usual experimental conditions encountered in materials processing and sophisticated techniques must be used to make significant progress. A better understanding of plasma spraying is now emerging as shown by a recent review of the Russian work in this area [66]. A comprehensive review of the plasma spraying of ceramics has also been recently published [185].

4.1.1 Macroscopic Properties of Coatings

The properties required of plasma sprayed coatings may vary considerably depending upon the application; for example, low porosity is necessary for wear resistance and corrosion protection but very porous deposit may be more suitable for thermal barriers. The physical properties of a coating of a given material will largely depend upon the spraying conditions.

4.1.1.1 Structure

A plasma sprayed deposit consists of successive layers of material built up by the impact of molten droplets, projected at high velocity, which flatten against the substrate and solidify very rapidly. As emphasised by Rykalin [66], even with high powder feed rates the particles are deposited on already solidified layers. Heat transfer calculations show that freezing of the particles occurs in a few microseconds and that complete cooling amounts to perhaps 100 μsec. The zone of thermal effect in the underlying material is therefore quite small and temperature gradients reach 10^5 K cm^{-1}. The coating has a sandwich like structure as shown in Fig. 57 and its properties can be regarded as resulting from the deformation and solidification processes of individual particles, and/or their interaction on contact. The wetting and flow properties of the liquid droplets are of first importance since they will influence porosity within the coating and at the substrate interface.

The flow and solidification of molten particles on impact with a cold surface is a difficult problem to treat theoretically because of the interaction between heat transfer and crystal growth. The very high velocity at which the various processes occur also makes direct experimental observation extremely difficult.

Madejski [186,187] has carried out a theoretical treatment of the impact of a molten

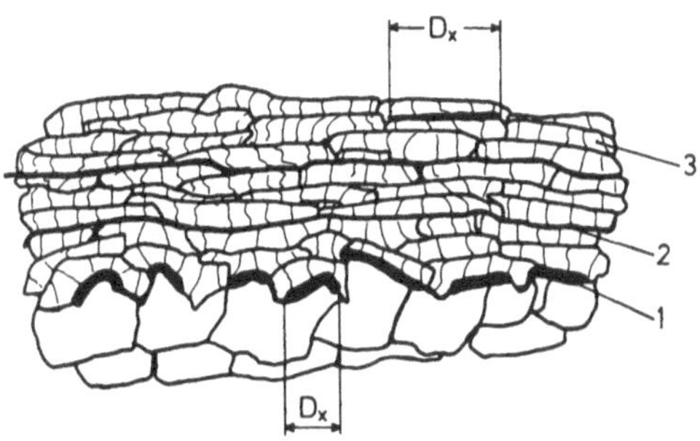

Fig. 57. Texture of plasma sprayed coating: (1) contact surface between the coating and substrate; (2) contact surface between layers; (3) contact surface between single particles. Reprinted from [66]

droplet on a cold substrate in which he has attempted to include surface tension, viscosity and crystallisation effects. Simplifying his treatment by neglecting the influence of surface tension, which is small for plasma spraying conditions, and assuming that the liquid droplet flattens before solidification occurs, gives for the degree of flattening (ratio of the diameter of flattened disc (D) to diameter of the initial drop (d))

$$D/d = 1.29\left(\frac{\varrho v d}{\mu}\right)^{0.2}$$

where ϱ is the liquid density, μ the liquid viscosity, v the droplet impact velocity.

Substituting appropriate values for Al_2O_3, in the velocity range 100–400 m/s gives D/d = 3–6, which seems to be in reasonable agreement with various experimental observations [109,188,189]. This suggests that, at least for Al_2O_3 under the spraying conditions, flattening is completed before crystallisation becomes significant and the two processes can largely be treated separately. The analysis assumes that the particles wet the substrate and do not break up during flattening; however, there is some evidence that this is not true under certain conditions as shown by experiments in which the splash pattern of single particles has been studied [189,190]. It should be born in mind however that smooth substrates have been used in these studies and under normal spraying conditions the substrate is grit blasted and a rough surface persists during the deposition of subsequent layers.

Plasma spraying conditions are very similar to those occuring during splat quenching of metals which has been widely studied. The cooling rate has been of particular interest and various techniques have been used to estimate the thermal conditions as a molten particle spreads and solidifies [191]. The main results of this work show that the cooling rate is largely independent of substrate material and is controlled by the interface between substrate and particle if the flattened particle has a thickness of less than a few μm. The estimated heat transfer coefficient under these conditions is of the order of 10^5 W m^{-2} K^{-1}. Applying this result to the case of plasma sprayed Al_2O_3 gives a cooling rate of the order of 10^7 K/s and a freezing time of about 10 μs. Such an analysis also suggests that there is a large temperature gradient across the interface between the particle and substrate (or particle and previously frozen material) and the temperature of the underlying material therefore remains relatively low. This is particularly so if the substrate is cooled during spraying by an air blast to prevent heating by plasma gases and to remove the sensible heat and heat of fusion of the sprayed particles.

It may be seen that the structure of a coating will depend upon the velocity, temperature and size of the particles at the moment of impact. Analysis of the process is, however, made difficult because there will be a statistical distribution of each of these parameters and it is only possible to make some general theoretical predictions [109]. Such analysis are, however, of considerable importance for the interpretation of experimental studies of coating structure.

4.1.1.2 Coating Adhesion

The adhesion of a coating to the substrate is one of the most important properties of a plasma sprayed deposit but the factors controlling adhesion are not well understood and there is some controversy concerning the mechanisms involved. One point of view suggests that adhesion is controlled purely by mechanical factors [192], that is, interlocking of the coating with the rough substrate surface. Another is that chemical interaction occurs between particles and substrate [193]. Adhesion is however a complex property and neither chemical interaction nor mechanical locking is necessary for adhesion to occur. A surface reaction hypothesis [185,193] has been studied by Russian workers using model experiments with relatively large silver particles to test the hypothesis. They suggested that the strength of the bond developed between particle and substrate is increased with substrate temperature and that increased contact pressure arising from impact velocity reduces the effective activation energy for bond formation. It remains to be shown however, whether this hypothesis is appropriate for practical plasma sprayed coatings because of the very short time during which pressure is acting and, since the range over which chemical effects can occur in such short times is extremely small, the effect of the nature and thickness of oxide films on the substrate becomes very important. The latter effect may be illustrated by the observation that the strength of the adhesion bond is reduced with delay time between grit blasting and spraying [194].

The difficulty in establishing a "metallurgical" bond between coating and substrate is shown by experiments involving plasma spraying of tungsten on to polished tungsten substrates in an inert atmosphere [195]. It was found that negligible bonding occurred at room temperature and a bond involving epitaxial growth occurred only if the substrate was heated to temperature greater than approximately 1300 °C.

The formation of interaction layers between coating and substrate has been observed in the case of the spraying of nickel aluminide using powders in which each particle is a composite of Al and Ni [196]. In this case a strong exothermic reaction occurs between the two molten metals, which probably continues after the particle has contacted the substrate, giving localised high energy input. Unless there is an exothermic reaction between coating and substrate, or, if the temperature of the impacting particles is very much higher than the substrate so that local substrate melting occurs, it is difficult to see how an interaction layer of any significant depth can ge formed because of the very short time available. Even assuming values for diffusion coefficient (D) in the substrate of the order of that near the melting point ($= 10^{-9}$ cm^2/s) the diffusion depth (x), given by x = Dx t where t is time, is only of the order of 10 Å for a contact time of 10 μs. Some examples [197] quoted in the literature for the formation of intermetallic layers probably arise from exothermic reaction between the two metals at the interface.

It seems unlikely that interaction layers could be formed between a metal substrate and ceramic coating although it is conceivable that interaction would occur between sprayed droplets and the oxide layer present on any metal in the atmosphere.

This brief discussion can only give a superficial view of the problem of coating adhesion; however, there seems no doubt that much more work remains to be done before the mechanism is properly understood.

4.1.1.3 Internal Stresses in Coating

The residual stresses in coatings may be subdivided into microstresses within individual particles and macrostresses within the coating as a whole. Microstresses arise because of the restraint due to the thermal contraction of individual particles, as they cool in the solid state and by the underlying material which remains at a relatively constant temperature. These stresses will therefore depend largely upon the expansion coefficient of the coating material and the elastic constants of coating and substrate. They would also be expected to be influenced at the interface by the yield strength of the substrate, plastic deformation of which would allow stress relaxation, and also the effectiveness of the particle-substrate bond.

Macroscopic residual stresses will arise, after cooling of the coated structure to ambient temperatures, because of the difference in thermal expansion between coating and substrate and the presence of temperature gradients during coating formation [198]. High interfacial stresses may lead to peeling of the coating, particularly on smooth substrates, and high tensile stresses in the coating itself may lead to cracking. A residual compressive stress in ceramic coating could however be desirable as a method of increasing the fracture strength. To reduce residual stresses it is obviously necessary to reduce temperature gradients as much as possible and to keep the complete assembly at a low temperature by cooling the surface with an air blast. It is more difficult to reduce the stresses to a low value when producing coatings of low thermal conductivity, particularly as the coating thickness increases.

Measurement of residual stresses in plasma sprayed Al_2O_3 coatings on steel showed that they were compressive, with little change with thickness, whereas tensile stresses which increases with coating thickness were observed for nickel alloy coatings [199]. The residual stress of Mo coatings changed from tensile to compressive as the thickness was increased above 0.4 mm. Both micro and macro residual stresses would be expected to influence coating adhesion but little data on this topic seems to be available. An interesting possibility is the improvement of the integrity of ceramic coatings by establishing an optimum residual stress distribution by suitable control of spraying conditions.

4.1.1.4 Porosity

Although high porosity may be desirable for thermally protective coatings, porosity greatly reduces the strength of the coating but interconnected porosity is undesirable in coatings for oxidation or corrosion resistance. Porosity at the substrate interface will also greatly reduce coating adhesion. The nature of coating formation by impact and solidification of separate droplets necessarily results in some porosity which generally lies in the range 5–20% and depends both on spraying conditions and the material being sprayed. A material dependent effect may be seen from a classification given by Ingham [200]. For example TiO_2, Al_2O_3 – 50% TiO_2, Cr_2O_3 and NiO are included in a group in which porosities of less than 4% may be achieved. Spinel, mullite, Al_2O_3, ZrO_2 are included in a group in which porosities range between 4 and 8% and rare earth oxides and zircon in a group in which porosities range between 8 and 15%.

This variation with composition is perhaps associated with the wetting characteristics of the various materials and related to surface tension and contact angle. It is interesting to note that the surface tension of TiO_2 is about half that of Al_2O_3 at their respective melting points [201] and that the presence of a small amount ($\simeq 2.5\%$) of TiO_2 reduces the porosity of Al_2O_3 coatings. In general, the materials giving low porosities also gives the best adhesion to the substrate which would also be expected to arise from improved wetting.

For a given coating material the porosity is related to the particle size distribution of the powder and to the temperature and velocity of the particles on impact. Clearly the lowest prosity coatings would be expected for a stream of completely molten particles with high velocity. The role of velocity is demonstrated by experiments at Limoges [109] in which γ-Al_2O_3 particles with narrow size distribution (18 $\pm$ 4 μm) were sprayed using a N_2—H_2DC plasma jet.

A porosity of 23% was observed at a 75 mm torch-target distance but this was reduced to 11% at 50 mm, the distance which was found to correspond to the maximum particle velocity on the axis of the jet. The pore morphology was different in each case with much narrower interconnecting porosity in the low porosity coating. The effect of velocity is also shown by the radial variation of porosity in a deposit sprayed onto a fixed target. The particle velocity on the axis of the jet was observed to be 150 m/s greater than that at the periphery (Fig. 58), and the porosity of the deposit was observed to range from 10% on the axis of the jet to 22% at 15 mm radius.

The temperature, velocity and thermal conductivity of the gas in a plasma jet varies considerably from point to point and, since the trajectories of individual injected particles will differ, the spray stream produced will contain particles with a wide range of velocities and temperatures. In practice a considerable proportion may not be melted or be only partly melted [109, 124], or have low velocities and these will tend to reduce the quality of the coating. This effect can be reduced to some extent by a compressed air blast normal to the plasma jet, and close to the substrate, which removes a large proportion of the particles which have not mixed properly with the plasma and thus have low velocities [202].

There is little doubt that the lowest porosities are achieved with high particle velocity as shown by the low porosity coatings produced by the detonation method in which impact velocities of 750 m/s are obtained [203]. The use of high velocity plasma jets would seem desirable for the preparation of low porosity coatings but they must also be of high power to provide sufficiently rapid heat transfer to melt the particles. A practical limitation may be reached, however, with ceramic materials of low thermal conductivity because of surface vaporisation. This approach is being intro-

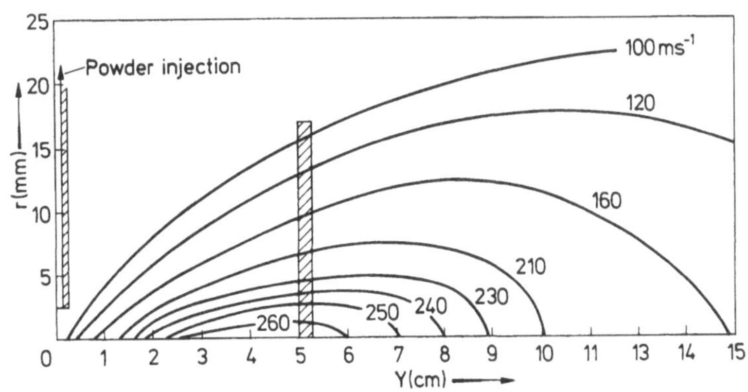

Fig. 58. Alumina particle velocity field in a DC nitrogen-hydrogen plasma jet. The substrate is located at 5 cm from the nozzle exit. Reprinted from [109]

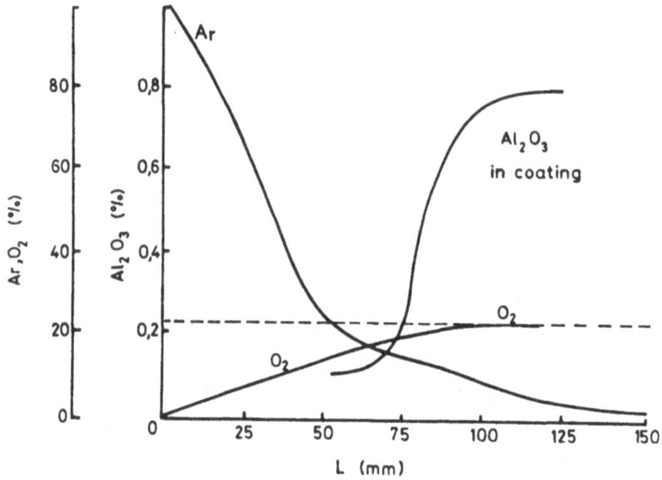

Fig. 59. Oxygen and argon content in an argon plasma jet and aluminium oxide content in the coating depending on the distance from the nozzle L. Reprinted from [66)]

duced commercially using jets with supersonic nozzles and considerably higher power inputs (80 kW) than generally used [204)].

4.1.1.5 Chemical Changes in Sprayed Particles

Chemical changes may occur in particles in a plasma jet by reactions with the gas. As we have already seen the chemical reaction is governed by diffusion of the plasma gas towards the condensed phase and by the inverse diffusion of gaseous product. This diffusion occurs at a rather low rate (10^{-3}–10^{-1} s) compared with the melting time (10^{-4}–10^{-3} s), especially when hydrogen or nitrogen is used in place of argon. So even with the rapid diffusion of air into a jet when spraying in the atmosphere (see Fig. 59 for an Ar Plasma [66)]), the proportion of Al_2O_3 in the aluminium sprayed coatings is only 0.1% at 50 mm and increases to 0.8% at 150 mm [66)]. Practically all of the particles are melted at a distance of 30 mm, and thus with a target at this distance, chemical transformation is negligible. Oxidation is however more of a problem with metals such as W, Mo, and Ti and inter-particle oxide may be detected in the coating [205)]. Although dissociation of Al_2O_3 has been observed in experiments using RF plasmas, the residence times are orders of magnitude greater than for plasma spraying and reaction with plasma gases does not seem to be a problem with oxides. A matter that does remain largely unresolved is the formation of gas bubbles in sprayed coatings and spheroïdized particles (see Sect. 4.3.). In the case of Al_2O_3 bubble formation has been ascribed to reactions with N_2 to give AlN in the liquid droplet with subsequent decomposition [189)]. Internal porosity is, however, also observed in Al_2O_3 spheroïdized in inert plasma gases [206)]. Reaction does become a problem with the spraying of carbides and the structure of carbide deposits is often extremely complex for this reason [207)].

4.1.1.6 Crystal Structure

An extremely rapid cooling rate from the liquid state is characteristic of plasma sprayed material and this may result in the suppression of crystallisation or the formation of metastable phases as observed under similar conditions in splat quenching. The classic example of this effect is provided by alumina, coatings of which consist predominately of one or more of the many metastable forms ((η, γ, δ, θ) rather than α-Al_2O_3 the only stable structure [188]. The formation of metastable phases may be explained on the basis of a lower activation energy for nucleation of the spinel type structure from the liquid when it is undercooled well below the equilibrium melting point as a result of rapid cooling [208]. The phase finally observed depends upon the rate of transformation from one metastable form to another and to α-Al_2O_3, and the cooling rate of the solid after solidification. This explains the observations that different metastable phases are observed depending upon the spraying conditions, size and shape of particles and the temperature of the substrate [209]. It is necessary to heat the substrate to above 1100 °C to obtain a coating consisting completely of α-Al_2O_3 [210] and to 1450 °C to obtain a well sintered deposit [211]. The α-Al_2O_3 present in conventional coatings (10–25%) is probably due to incomplete melting of the powder feed resulting in the presence of α-Al_2O_3 in the droplet in impact [188] which acts as nuclei for crystallisation. Although a deposit of metastable alumina may be heat treated to transform it to the more desirable α-Al_2O_3 form, for the preparation of bodies by spraying onto a removable core, the relatively large change in the true density from 3.6 for γ-Al_2O_3 to 4.0 for α-Al_2O_3 results in an increase in porosity and consequently a degradation in mechanical properties [212].

A further effect observed in some coatings is a preferred orientation of the crystals relative to the surface. This is observed in Cr_2O_3 deposits in which the c-axis of the hexagonal cell is preferentially oriented normal to the surface, apparently arising from the solidification process [213]. Since the coefficient of friction tends to be lower in the closely packed planes of a crystal [214] (basal plane of hexagonal crystals) this means that Cr_2O_3 deposits are favorably oriented to give minimum coefficient of friction. Sprayed chromia is therefore a useful material for low friction bearings and its high hardness makes it useful under conditions in which abrasive wear is a problem. Other hexagonal hard compounds such as W_2C, CoMo and Co_3W also provide coatings with low coefficient of friction, the value of the friction coefficient being particulary low (0.15) for CoMo [213]. The differences in coefficient between the various materials appears to be related to the details of the deformation mechanisms by slip [215].

4.1.2 Sprayed Materials

In principle any material with a melting point a few hundred degrees lower than the boiling point or dissociation temperature may be plasma sprayed. A review of the most commonly used ceramic coatings is given by Boch [185].

4.1.2.1 Oxides (See Table 13 from Ref. [216])

Oxides have excellent stability, good mechanical properties (hardness, wear resistance, strength) and thermal properties and are generally available at a low cost. The most commonly used oxides are:

Table 13. Properties of flame sprayed refractory oxides

Types / Properties	Aluminium oxide	Zirconium oxide	Zirconium silicate	Aluminium silicate	Titanium dioxide	Chronium oxide	Iron titanate	Rare earths (50% cerium oxide)
Application properties								
Deposition rate, sq ft/hr/10 mils	15–16	4–40	—	16	10	—	10	8
Thickness range, mils ($m \times 10^{-4}$)	5–50[a] (1.27–12.7)	5–50[b] (1.27–12.7)	5–50 (1.27–12.7)	—	—	—	—	—
Finish after polishing, μ in. rms ($m \times 10^{-8}$)	2–50[c] (5.08–127)	15–50[c] (38–127)	30–50 (76–127)	52–77 (132–195)	6–43 (15–109)	—	7–52 (17–132)	32–48 (81–121)
Thermal properties								
Melting point, F (K)	3600–3700 (2255–2300)	4500–4600 (2755–2800)	3000 (1922)	3200 (2035)	3200 (2035)	3000 (1922)	2500 (1644)	—
Coef of ther exp. 10^{-4} per °F (m/m/K)	4.0–4.3[c] (7.2–7.74)	5.4–6.4[c] (9.72–11.52)	4.2 (7.56)	2.5 (4.5)	4 (7.2)	5 (9)	—	5 (9)
Ther cond, Btu/hr/sq/ft/°F/in. (W/m — K)	19–20 (2.74–2.9)	7–8 (1.0–1.15)	15 (2.16)	25 (3.6)	45 (6.48)	18 (2.6)	—	20 (2.9)
Ther shock resistance	Good	Very good	Good	—	—	Moderate	—	—
Other properties								
Dielectric strength, v/mil	200	100[d]	Noncond.	—	Conductive	Noncond.	Conductive	—
Vickers hardness (DPH, 25 gm load)	1400	400	1000 Knoop	650	1500	1900 Knoop	700	600
Porosity %	8–12	8–12	8–12	—	—	4	—	—
Density, lb/cu in. (kg/m^3)	0.12 (3321)	0.19 (5259)	0.14 (3875)	—	—	0.19 (5259)	—	—

[a] Thickness up to 0.125 in. (3.2 mm) can be produced for special applications; [b] Thickness over 0.10 in (2.5 mm) can be produced for special applications; [c] Varies with method of applications; [d] Nonconductive at room temperature conductivity increases rapidly above 2190 F (1475 K)

Al_2O_3 used as a heat barrier and for wear resistance. With a medium value (8×10^{-6} m/mK) for thermal expansion coefficient, it offers good adhesion on metal substrates but is rather sensitive to thermal shock. The main disadvantage for more general use is its reactivity with many molten salts and other metal oxides and its relatively high porosity. Al_2O_3 — alloys may be used to reduce the porosity whilst retaining the wear resistant properties. For example TiO_2 is commonly added in the range 2–50%. Mixture of Al_2O_3 with zircon offers good corrosion resistance.
ZrO_2 is used to protect substrates against oxidation or corrosion and as a thermal barrier. It is necessary to stablize ZrO_2 e.g. by additions of CaO or Y_2O_3, to avoid the large volume change which occurs on the transition to the monoclinic form.
Very hard Cr_2O_3 deposits ($\simeq$1900 Vickers hardness) with excellent wear resistance and low friction coefficient are obtained by plasma spraying. The deposits have low porosity and may be diamond ground to an excellent surface finish.
TiO_2 coatings exhibit excellent adhesion to the substrate, low porosity and are capable of providing the best surface finish available from ceramic deposits [200)].

4.1.2.2 Carbides (See Table 14 from Ref. [216)])

The high hardness of tungsten carbides, titanium carbides and chromiun carbides make them excellent coatings for wear resistant application. They are generally sprayed with a metal binder such as Co, Ti, Cr and a compromise must be reached between wear resistance and other properties which depend upon the percentage metal. Increasing the bond metal content decreases the wear resistance but improves the mechanical properties and resistance to thermal shock. Chemical changes such as carbon loss, and physical changes occur during spraying; the coating microstructures are quite complex and their properties are related to the spraying process parameters.

4.1.2.3 Borides and Silicides

Borides provide hard wear resistant coatings with a low reactivity for molten metals. The borides high neutron capture cross section also makes them useful in certain nuclear application.

Hard wear resistance coatings are also produced by sprayed silicides which are useful in air at high temperatures because of the formation of a protective silica skin.

4.1.2.4 Nickel Aluminide

A special position amongst plasma sprayed material is occupied by nickel aluminide coatings produced by spraying particles consisting of a layer of Ni over an Al core. As previously quoted these two metals react exothermically when melted and form a coating consisting of Ni_3Al which is very hard and adheres very well to metal substrates. This probably arises from reaction at the interface resulting from the exothermic reactions which continue after impact. Nickel aluminide coatings are mainly used as intermediate bonding layer between metal substrates and coatings such as Al_2O_3 and ZrO_2.

Table 14. Properties of flame sprayed carbides [216)]

Coating	Tungsten carbides			25% WC + 7% Ni + 85% chromium carbide mixed W-Cr carbides (Cr_3C_2) + 15% Ni-Cr	
Properties	WC + 11% Co	WC + 13% Co	WC + 15% Co		
Mechanical properties					
Vickers hardness	1300	1150	1050	1000–1200	1000–1200
Mod of rupture, psi	80,000	90,000	100,000	40,000	75,000
(MPa)	(552)	(620)	(689)	(276)	(517)
Mod of elast. 10^6 psi	30	31	31	17	22
(MPa × 10^5)	(2.07)	(2.13)	(2.13)	(1.17)	(1.51)
Porosity, %	1.0	1.0	1.0	0.5	0.5
Thermal properties					
Max cont svc temp, F	1000	1000	1000	1400	1800
(K)	(811)	(811)	(811)	(1033)	(1255)
Coef of ther exp. 10^{-6} per °F	4 (70.1000 F)	4.5 (70.1000 F)	4.7	4.6 (70.1400 F)	6.4 (701.800)
(m/m/K)	(7.2; 294–811 K)	(8.1; 294–811 K)	(8.46)	(8.28; 294.1033 K)	(11.5; 294.1255 K)
Spec.ht, Btu/lb. °F	0.048	0.054	0.056	0.07	0.127
(J/kg · K)	(200)	(226)	(234)	(293)	(531)
Ther.cond.Btu/hr/sq ft/°F/ft	5.3 (70.500 F)	5.3 (70.500 F)	5.3 (70.500 F)	3.8 (500 F)	4.3 (500 F)
(W/m · K)	(9.2; 294.533 K)	(9.2; 294.533 K)	(9.2; 294.533 K)	(6.6; 533 K)	(7.4; 533 K)
Major characteristics	Extreme wear resistance	Excellent wear resistance plus increased resistance to mechanical shock	Excellent wear resistance plus greater resistance to mechanical and thermal shock	Excellent wear resistance at high temperatures; improves corrosion resistance	Good wear resistance at high temperatures or in corrosive media; resists flame impingment

4.1.2.5 Composite Materials

Composite coatings are commonly used [200], for example, a MoO or Ni_3Al intermediate layer with ZrO_2 or Al_2O_3. However, it is also possible to prepare fibre reinforced composite materials by plasma spraying, which have properties considerably exceeding those of conventional materials [217, 218].

4.2 Spheroidization

As previously quoted any material with a melting point well below its vaporisation or decomposition temperature can be melted in a plasma jet, and perhaps to some extent vaporised, depending upon its size, physical properties, residence time in the plasma and rate of heat transfer from plasma to particle. Melting of a particle results in the formation of a spherical drop under the action of the surface tension forces and this shape is usually retained by the solid particle after solidification, thus providing the name to the process [219-221]. This treatment may be used simply to give a spherical particle shape for particular application, such as Fe_3O_4 for photocopying and UO_2 for dispensed nuclear fuels. Structural changes resulting from melting and solidification may also be important. For example, with the Ionarc furnace (Fig. 44) the dissociation of Zircon to Zr_2O_2 and SiO_2 allows the preferential dissolution of SiO_2 and the preparation of ZrO_2 [222]. Spheroidization of quartz, clays, kyanite, mullite and other alumino silicates results in the formation of glasses of high melting point [223-225].

The problems associated with the spheroidization process are similar to those of spraying because in both cases it is necessary to obtain rapid melting without evaporation and often, without chemical change. The melting process may however provide a purification effect by evaporation of some impurities, for example Fe, Mo, Na from Al_2O_3.

It may be possible to control the composition of particles by spraying in a controlled atmosphere in a chamber with diameter of 2–10 times that of the plasma jet. For example, the decarburisation of carbide particles may be prevented by cooling them in chamber into which hydrocarbons are introduced and by adding carbon in the form of soot to the starting powders [128]. Structural changes, such as changes in crystalline structure and crystal size, formation of additional phases or glass formation are certain to occur during the melting-solidification process. Dendritic crystal growth during solidification may lead to a roughened surface of the particle [226, 227] or internal shrinkage porosity in the densities of liquid and solid are significantly different. It is also possible for the particle surface to be impaired by the formation of various deposits during cooling [228].

Plasma spheroidization began in 1961 on a laboratory scale using a 25 kW R.F. plasma torch [229] and has since expanded to an industrial level. R.F. plasma torches are generally used for spheroidization of ceramic particles (relatively long residence time and no pollution by electrodes) when high purity is required. Direct or alternative current arc plasma jet gives rise to limitations in the particle size that may be treated because of the short residence time. Plasma furnaces of the B type, (fluid convective cathodes or fluidised bed), which give longer residence times maybe more suitable for large scale spheroidization at larger particle sizes.

Spheroidised particles may also be prepared by atomisation of rods or wires fed into a plasma. An analysis [66] of the forces acting on the molten drops at the ends of the rod shows that the diameter of the produced particles can be expressed by:

$$d_r = \frac{3.35\,d_c^2}{Q(1 + 3.67 \times 10^{-3}\,T} \cdot \frac{d_w \sigma}{\varrho}$$

where d_c is the plasma torch nozzle diameter (m), Q the gas flow rate ($m^3\ sec^{-1}$), d_w wire diameter (m), σ surface tension (Jm^{-2}), ϱ plasma density ($kg\ m^{-1}$), T temperature of plasma (°C).

The higher the plasma temperature, the smaller is the particle diameter. Using an Al_2O_3 feed rod and a hydrogen plasma, particles of 40 μm were obtained. The heat transfer can be improved if the arc is transferred to the wire but this is not possible with non conducting ceramics.

A large number of references to spheroidization have been provided in review articles by Waldie [220] and more recently Hamblyn [219] and Rykalin [66].

An example of large scale industrial application is the spheroidization of magnetite (125 μm) for photocopying applications, in an A.C. air plasma heater of 600 kW with a power consumption of 2 kWh/kg [230].

Another important industrial application of spheroidization of ceramic interest is the dissociation of zircon sands developed by Ionarc Tafa using a fluid convective cathode plasma furnace (see Sect. 3.3.1 and Fig. 44) operating at 300 kW with feed rates of 140–270 kg/h depending upon the degree of conversion required [222]. A similar device has been studied by Sayce using 130 kW but with three plasma torches instead of three carbon electrodes [162]. In these processes the particles are either dissociated into ZrO_2 and SiO_2 without melting (or with partial melting) or melted to form ZrO_2 crystallites in a SiO_2 glass matrix [231]. The major purpose of this treatment is to produce a structure from which SiO_2 may be readily leached with mild acqueous sodium hydroxide leaving porous ZrO_2 particles which may be used in ceramic applications or as a source of Zr metal. The dissociated zircon has potential applications as a ceramic material in glasses [232] or in the manufacture of ceramics with high thermal shock resistance by sintering to give a microstructure consisting of particles of ZrO_2 in a $ZrSiO_4$ matrix [233].

Another possible ceramic application of plasma spheroidization is the preparation of glass powders for use in manufacture of high temperature glass ceramics [224].

4.3 Vaporisation — Condensation

Vaporisation of materials in plasmas has been extensively studied and is becoming of increasing interest from the commercial point of view, especially for the preparation of ultrafine powders characterised by enhanced chemical activity, and also as a means of separating valuable products from otherwise intractable materials [234]. From the ceramics point of view, plasma processing offers the possibility of preparing powders for subsequent sintering which either cannot be prepared in any other way, or have superior properties to powders prepared by conventional means. The process of vaporisation and subsequent condensation to useful products has been extensively treated by Sayce [183] in a recent paper and we shall use many of his conclusions.

4.3.1 Vaporisation

4.3.1.1 General Remarks

The vaporisation of a particle at atmospheric pressure requires a large amount of energy to supply the difference in enthalpy between the solid and vapour phases. Furthermore as vaporisation proceeds, the heat must be carried through the expanding vapour film on the particle surface. For example, a 200 μm particle of silica will produce a cloud of vapour approximately 5 mm diameter at 3000 K (assuming $SiO + {}^1/_2\, O_2$ and uniform temperatures across the cloud) [185]. The vaporisation process therefore requires conditions under which high heat transfer rates may be achieved and the plasma furnaces used are usually of the C and D type: rotating furnaces or arc transferred to the materials being vaporised (see Sect. 3.3.).

As we have seen a number of theoretical models have been developed for the vaporisation process [235–237] but an interesting result emphasised by Capitelli [238] is the probable cause for the increase of heat transfer. Due to the higher temperatures of the electrons (non LTE plasma), this effect would be specially important when the material being vaporised is used as an electrode.

4.3.1.2 Materials Treated

One of the most extensively treated materials is silica with the aim of preparing ultrafine silica powder which has a large number of industrial applications and is used at a rate of about ten thousand tons per year. A wide variety of technique have been used:

(I) D.C. holow cylindrical electrodes (see Fig. 54) [239], with a thermal efficiency of 25–53 kWh/kg
(II) Horizontal rotating drum furnace heated by a single plasma torch [172] (10 kWh/kg)or carbon arc (9.3 kWh/kg) [183] (see Fig.56)
(III) High intensity arc [240] (9–18 kWh/kg)
(IV) Fluid convective cathode [183] (14 kWh/kg)
(V) Three phase plasma fluidised bed [241, 164] (see Fig. 45).

In all cases the energy consumption is reduced if silica is treated in the presence of reducing agents such as H or C. Then SiO_2 is vaporised as SiO which is subsequently reoxidised to SiO_2 during condensation.

Other oxides may also be treated in this way. For example, Mg, Zn, Fe, Sb oxides have been treated using the FCC furnace of Sheer [242]. Extremely fine Al_2O_3 particles have been prepared by vaporisation using the rotating drum furnace of Sayce [243, 244]. With the same furnace fuming of the low tin slags has also been investigated [245]. With an energy consumption of 4.5 kWh/kg, 80% of the tin content has been removed from a 3.4% tin slag.

4.3.2 Condensation

The preparation of solid materials by condensation of vapours formed by plasma reaction can be carried out in several ways; chemical vapor deposition on a substrate (this process is very important for the preparation of thin films in low pressure plasma); condensation to the liquid phase on a cooled solid to produce, for example, pieces of high purity SiO_2 glass with low OH content for use in optical fibers; condensation of the vapour to obtain very small particles. We will deal exclusively with the latter process which is of considerable practical interest in the preparation of materials for ceramic applications.

4.3.2.1 Nucleation

The temperature attainable in plasma reactions are such that any material may be transformed into the vapour phase, either by vaporisation of the solid as discussed above, or by high temperature reaction between a reactant vapour and the plasma gas. Examples of the latter method are the preparation of oxides (SiO_2, TiO_2, Al_2O_3) by the oxidation of the respective halides. The preparation of TiO_2 as a pigment is an important industrial example of this application of plasmas in technology and large plants are in operation using R.F. [128)] or arc plasma devices of high power [246)]. Carbides, nitrides and other compounds may also be prepared by similar methods. For example SiC by reaction of $SiCl_4$ and CH_4 [247)], TiC by reaction of $TiCl_4$ and methane [128)], nitrides by reaction of metal and N_2, or halide and N_2.

It is found that the presence of reactants such as metal halides in the plasma causes problems with stability in the case of R.F. devices and corrosion in electrode plasmas. However, the extreme plasma temperatures are not necessary and satisfactory results are obtained by injecting reactants into the plasma tail flame as demonstrated by Audsley and Bayliss for the production of SiO_2 by reaction of $SiCl_4$ and oxygen using an RF plasma [248)].

Whatever the source of the vapour phase, the subsequent condensation to particles passes through several stages, the first of which is nucleation. As the vapour cools in the exhaust of the plasma reactor, a stage is reached at which the vapour pressure exceeds the equilibrium vapour pressure of a condensed phase in the system and the formation of this phase then becomes thermodynamically possible. However, a certain degree of undercooling below this temperature (or a certain super saturation of the vapour) is necessary before nucleation of the condensed phase occurs at a significant rate. This is because activation energy is required for the formation of the new interface between the vapour and condensed phases. The classical nucleation rate equation gives:

$$I = A(T) \exp\left(-\frac{\Delta G_a}{kT}\right)$$

where ΔG_a is the activation energy for nucleation, I is nucleation rate and kT has its usual meaning. The value of the nucleation rate is completely dominated by the exponential term because ΔG_a depends sharply on supersaturation:

$$\Delta G_a = \gamma^3/(\ln p/p_0)^2$$

where γ is interfacial energy and p/p_0 is the supersaturation. The result is that there is a critical supersaturation at which the nucleation rate increases from a negligible value to an extremely large one and nuclei of the condensed phase are formed. The critical size of these nuclei may only be of the order of nanometers. In practice charged particles may reduce the supersaturation necessary to cause nucleation and in the case of plasma processing, nucleation probably occurs at a supersaturation of 2–3 [249]. This means that as the vapour is cooled nucleation occurs at a temperature close to that at which the partial pressure of the condensed phase is equal to the equilibirum partial pressure. Nucleation will increase at an enormous rate as the vapour cools further and at the same time the nuclei will increase in size by surface reaction with the vapour, and if they are liquid by coalescence of the particles as they collide. This process has been treated in detail by Ulrich [249] and Cozzi and Cadoran [250] for the condensation of SiO_2 and TiO_2 from flames. Their results, although differing in the relative importance of the surface reaction contribution to particle growth, emphasize the rapidity of nucleation and the importance of coalescence in determining the final particle size. The particle size obtained will thus be determined by the initial concentration of the oxide in the vapour phase and the opportunity for coalescence, that is the cooling rate and the temperature range between nucleation and solidification of the particles. Results obtained for condensation of $Al_2O_3 - SiO_2$ [251] and $Al_2O_3 - TiO_2$ [252] powders from a high frequency plasma are in good agreement with this treatment. The conditions in the gas stream in these cases were such that complete condensation occured well above the freezing point of the product and the powder formed consisted of spherical particles 0.1–0.3 μm diameter (Fig. 60). Condensation of silica is a special case because liquid silica is very viscous and crystallisation does not occur on cooling. It would therefore be expected that below a certain temperature particles would weld on contact but would not flow together to form spherical particles. The rate of coalescence of particles may be estimated from equations developed for sintering by viscous flow [253]. Applying this to the case of SiO_2 particles 10 nm diameter shows that the time of coalescence would be approximately 3 ms at a viscosity of 10^6 poise, the value at $\simeq$2200 °C.

Since the cooling rate of particle in a plasma stream would range from 10^4 to 10^6 K, coalescence to form spherical particles would be expected to cease at temperatures below approximately 2200 °C. Condensation below this temperature would therefore be expected to result in the formation of chains of very small particles. This is in agreement with the observation that spherical SiO_2 particles

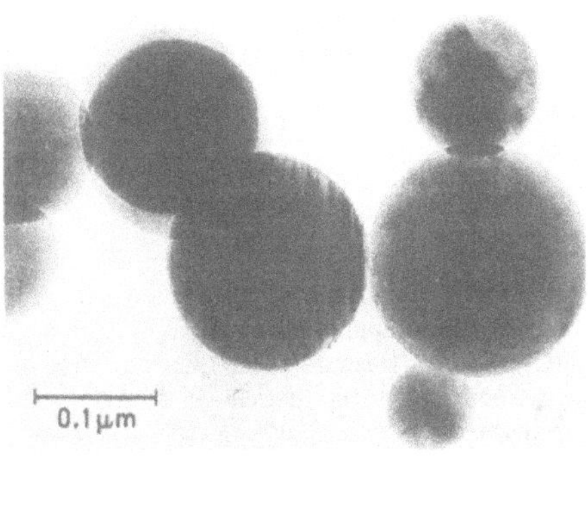

Fig. 60. Transmission electron photomicrograph of particles of pure $\delta \cdot Al_2O_3$ prepared by condensation from RF plasma. Reprinted from [252]

0.2 µm in diameter are produced by oxidation of $SiCl_4$ in an RF plasma under conditions in which condensation occurs at $\simeq 2700$ °C [248, 251], whereas chains of particle $\simeq 10$ nm diameter are produced by flame oxidation of $SiCl_4$ below $\simeq 2000$ °C. Sayce [183] observed that large surface area silica powders with chain like configuration were produced if the vapour from the vaporisation furnace was quenched to give a low effective pressure of SiO_2 and hence low condensation temperature which also agrees with the above reasoning. The rather unusual proportion of active silica powders depend upon their larger surface area and chain like structure but also on hydroxylation of their surface and the presence of H_2O in the system is said to be important to achieve optimum properties [183].

The nature of the condensation process in those materials which condense to liquid droplets rather than solid particles means that it is possible to co-condense two or more components to give liquid solution droplets. Coalescence during cooling contributes to a homogeneous composition for the powder if the flow rates of the feed vapours are maintained constant. The conditions under which homogeneous solutions may be formed have been discussed by McPherson [254] and examples quoted of $Al_2O_3 - SiO_2$, $Al_2O_3 - Cr_2O_3$, $Al_2O_3 - TiO_2$, $Al_2O_3 - FeO_3$ prepared by oxidation of mixed halides in an RF plasma tail flame. An interesting case is that of $Al_2O_3 - SnO_2$ in which Al_2O_3 droplets solidify before SnO_2 formation becomes thermodynamically possible, thus resulting in powders consisting of SnO_2 crystals heterogeneously nucleated on Al_2O_3 spheres.

Depending upon the vapour pressure-temperature relationship of the condensing material, its freezing point and effective concentrations in the gas phase, condensation may occur directly to the solid rather than the liquid phase. In this case the particle size of the powder produced will depend upon the crystal growth rate from the vapour since coalescence will not be effective, and the particles will tend to have regular crystallographic facets. MgO condensed from an arc, for example, consists of small cubic crystals [255].

4.3.2.2 Structure of Powders

The solidification of liquid droplets suspended in a cooling gas stream will be different to that observed under most normal conditions because they will tend to undercool well below the equilibrium melting point before crystallisation is nucleated.

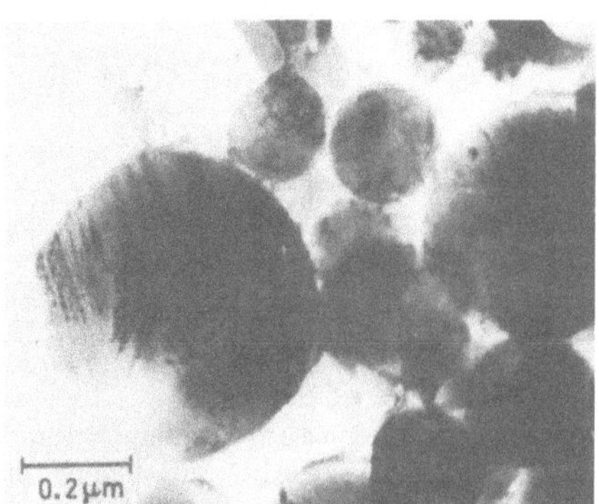

Fig. 61. Transmission electron photomicrograph of thin foil prepared from Al_2O_3 — 20 wt% TiO_2 powder treated in RF plasma, showing substructure of rutile particles in $\delta \cdot Al_2O_3$. Reprinted from [252]

It is well established that most metallic and ionic materials undercool to approximately 80% of the equilibrium melting point before crystallisation commences. The cooling rate is also large ranging form $10^4 - 10^6$ K/s. These effects may result in the formation of metastable phases as discussed in detail by McPherson [208] for alumina. The crystalline structure formed may be influenced by a second component. For example Al_2O_3 favours the formation of rutile form of TiO_2 rather than anastase [252].

The addition of Cr_2O_3 or of SiO_2 results in the formation of the metastable $\theta-Al_2O_3$ [256] or $\gamma-Al_2O_3$ [251] forms respectively, whereas pure Al_2O_3 is observed to be $\delta-Al_2O_3$ in powders prepared by oxidation of halides in an R.F. oxygen plasma. They may also result in the formation of extended ranges of solid solubility, new metastable phases, or structures consisting of an extremely fine dispersion of two crystalline phases as observed in co-condensed powders such a $Al_2O_3-TiO_2$ [252] (Fig. 61).

The rapid cooling rate and small particle size of powders condensed from a plasma may also result in suppression of crystallisation either completely or partly. The powder then consists of an amorphous phase as observed in the case of the $Al_2O_3-SiO_2$ system in which completely amorphous particles may be prepared with Al_2O_3 contents up to $\simeq 80$ wt · % [251]. An analysis of the nucleation kinetics for this system gives good agreement with the experimental observations and suggests that viscosity of the liquid is the important parameter at low Al_2O_3 contents but that changes in the interfacial energy between liquid and crystal become predominant as the composition approaches pure Al_2O_3. The importance of particle size in addition to cooling rate as a factor controlling glass formation is also emphasized [251]. Although crystallisation is suppressed the cooling rate may not be sufficiently high to prevent phase separation in the amorphous phase by spinodal decomposition to give particles which consist of an extremely fine structure of two glasses. This is observed for example at SiO_2 — 37 wt · % Al_2O_3 [257] and possibly also in powders containing $\simeq 80$ wt · % Al_2O_3 [258]. In the latter case a structure is produced which is not possible by very rapid quenching of larger particles (20 μm) because the cooling rate to suppress crystallisation is also high enough to suppress glass phase separation. For particles cooling in the tail flame of an RF plasma, the cooling rate is sufficient to suppress crystallisation because of their small size, but is not sufficiently fast to suppress phase separation. It may therfore be seen that there are great possibilities for the preparation of very fine ceramic powders with unusual structures.

Although plasma prepared powders could be expected to have interesting applications as ceramic raw materials, little has been published on their sintering characteristics. Experiments conducted using plasma prepared $\delta-Al_2O_3$ have given disappointing results which can be explained by the removal of grain boundaries between particles of the compact during the transformation to $\alpha-Al_2O_3$ [259]. Sintering of $\gamma-Al_2O_3$ containg SiO_2 is negligible up to $\simeq 1500$ °C and the transformation to $\alpha-Al_2O_3$ is suppressed. This is of interest for catalyst applications in which it is desirable to maintain activity at high temperatures by retaining a large surface area and at the same time reducing transformation to the less active $\alpha-Al_2O_3$ form [260]. More interesting results from the ceramic point of view are obtained during the sintering of SiO_2 — 37% Al_2O_3 plasma prepared powders in which dense products with extremely fine crystal size may be prepared [257]. A further possible application of plasma prepared materials is the simultaneous hot pressing and crystallisation of high melting point glasses. Some experiments using a mullite glass have shown that the process is possible but the small temperature interval between the glass transition and crystallisation temperatures limits the densification obtainable. Addition of 5% NiO increases the density obtainable, however the maximum density achieved to date has been limited to $\simeq 80\%$ [224].

5 Ceramic Synthesis in Thermal Plasmas

5.1 General Considerations

The use of thermal plasmas in chemical synthesis is now well established [154, 155] and among all the possible syntheses those of ceramic materials are particularly interesting. Numerous investigations have been published on plasma ceramics synthesis since the experiments of Stokes et al. [261] and it is impossible to give a detailed review of them in the present paper. For this reason we have preferred to limit ourselves to two types of ceramic materials: nitrides and carbides. Indeed sufficient results have been obtained for these materials to undertake a review providing significant conclusions concerning reaction mechanisms and the future of the sythesis of ceramic materials in thermal plasmas. We will try, when describing the different mechanisms, to emphasize the role of the various aspects considered in the preceding sections:

high temperature chemical equilibrium,
thermodynamic properties at high temperature,
heat and mass transfer in gas-solid systems,
condensation.

In particular when proceeding to a synthesis from solid, or powdered reactants fed into the plasma, it is necessary to carefully study the preceding points applied to the materials of interest. Unfortunately many experimental studies have been carried out with a very poor theoretical background and this has led to experimental results which provide little information concerning feasibility, and possible future improvements of the process under consideration. We believe that many studies need to be repeated with a greater number of measurements supported by relevant mathematical modeling. In this section we do not pretend to give all the necessary information to take the "right way", but we would like to make researchers aware of the problems with some particular examples, especially in the synthesis of nitrides.

5.2 Nitride Synthesis

Nitride synthesis is a very well known reaction in materials science which has been solved satisfactorily in most cases with classical methods. Generally one applies the chemical vapor deposition or works with powders reacting at high temperature ($T \sim 1500$ °K) in nitrogen or ammonia. Unfortunately the reaction times are very

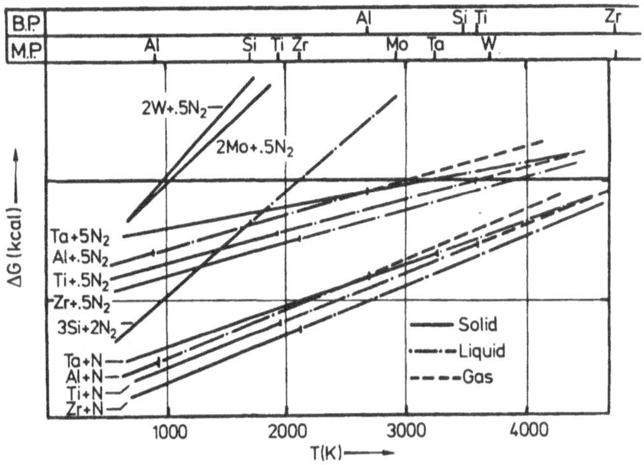

Fig. 62. Stability of nitrides at one atm. Reprinted from [290)]

long, and it is difficult to obtain very fine powders. The aim of the plasma chemist is therefore to reduce significantly the reaction time and to properly condense the products in order to obtain very fine powders. These objectives seem to be well fitted to plasma chemical synthesis because: the reactions take place at the high temperature of the plasma and the high quenching rates are favourable to the formation of fine powders.

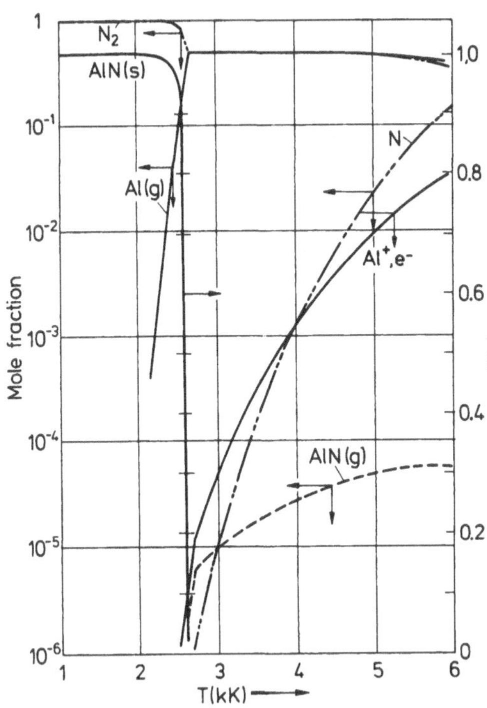

Fig. 63. Composition of the system Al, N_2 at one atm. Reprinted from [6)]

Unfortunately in real situation the reactions are not carried to completion. This is why one has to solve the problem of chemical reactivity before determining the appropriate quenching rate. These remarks also apply to carbide synthesis.

Matsumoto published in 1969 [290] a review of his work on nitrides synthesis. In the present paper we have summarized in table form (Table 15) the main experimental results that have been obtained up to now in this field.

5.2.1 Thermodynamical Considerations

The nitrogen plasma properties have been described in part 2. Let us consider now the thermochemistry of the mixtures (Metal—N_2). Figure 62 is a diagram of the free energy change versus temperature for the formation of nitrides from elements. (Some of these data come from the review of Matsumoto.) Except for W, Mo, Si, all nitrides become unstable only above 3000 K. Consequently reaction in the plasma has to proceed below this temperature limit. If one starts with powder materials, the particles have to be heated up to 2000 K, react and condense. (Obviously the problem is completely different if one starts with gaseous reagents.) Moreover, when using solid feed materials, the reaction mechanisms are different if the nitride is stable above or below the boiling point of the reactant. Figure 62 gives the melting and boiling points of the materials of interest. Consider for example the cases of the synthesis of aluminium and of silicon nitrides.

Aluminium nitride is stable with respect to the element above the boiling point of aluminium. Computed equilibrium compositions for the system Al—N_2 are shown in Fig. 63. The free energy changes versus temperature for this system are given in Figure 64 where it will be seen that the conversion of aluminium into aluminium nitride is completed up to 2700 K. At higher temperatures, the nitride decomposes to give gaseous aluminium and N_2 in equilibrium with N. The reaction Al + N →

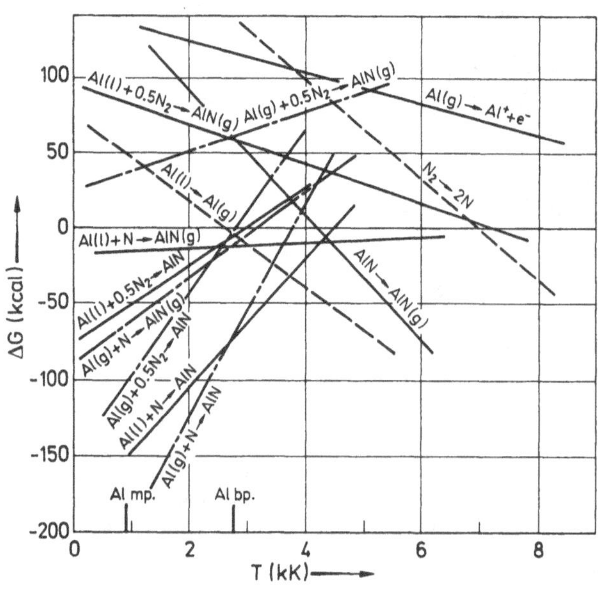

Fig. 64. ΔG changes versus temperature at one atm. for AlN synthesis. Reprinted from [6]

Table 15. Nitrides Synthesis

Nitride (case)	Reactants, Carrier gas	Plasma gas (l/min)	I (A)	V_a (V)	P (kW)	Quenching techniques Form of final products	Reaction yield	Ref. (year)
TiN (1)	Ti powders (200 mesh) (0.5 g/min) N_2 1 l/min Tubular injector ring below plasma generator	N_2 5			12	No information about quenching Finely divided powder (.75 to 7.5 μm)	100% TiN	261) 262) 263) (1960) (1964) (1965)
(2)	Solid cylinder of TiH_2 in the plasma stream.	N_2					100% TiN	264) (1970) (1971)
(3)	Densified, plasma prepared, TiB_2	N_2					Final product contains TiB_2 impurities	
(4)	Cylinder $2\,TiO_2 + B_4C + 3\,C$	N_2					TiN + T_iB_2 as a major impurity	
(5)	Ti metal button (10 g) in a water cooled copper crucible. Transfered arc on the crucible	N_2 1.5–4	60–120		3–7	Diffusion of nitrogen in the liquid metal. Powders are collected on the cold walls		265) (1968)
(6)	Ditto	do	do		do	Jv. the authors have determined diffusion coefficients of nitrogen in molten Ti A comparision with diffusion theory is proposed		(1973)
(7)	Ti powders (500 mesh) (Ar + N_2) (11, 11 l/min)	Ar 9	HF Plasma 4 MHz		5–35	Products are condensed on the inner walls of water cooled pyrex glass cylinder	100% TiN (non-stoichiometric)	267) (1977)

From powders or solid

Table 15. (continued)

Nitride (case)		Reactants, Carrier gas	Plasma gas (l/min)	I (A)	V_a (V)	P (kW)	Quenching techniques Form of final products	Reaction yield	Ref. (year)
(8)		Ti metal cylinder in a water cooled copper crucible. Non transfered arc.	N_2 4	70	150	10.5	Products are condensed on the cold walls of the reactor	85% after 20 min. Crucible at 2 cm from nozzle exit.	[268] (1968)
(9)	From $TiCl_4$	$TiCl_4$ (gas)	$64 N_2 + 36 Ar$ 24					40 l ($TiCl_4$) per g (TiN) 7%	[269] (1964)
(10)	From $TiCl_4$	$TiCl_4$ (gas) in $Ar + H_2$ mixture	N_2				Cooled quenching tube after nozzle exit.	7.09 l ($TiCl_4$) per g (TiN) 42.1% 10.2 kWh/g (TiN)	[270] [271] (1969) (1970)
ZrN									
(11)		ZrH_2 in the plasma stream.	N_2					Pure ZrN; 100%	[264] (1970)
(12)		ZrB_2 in the plasma stream	N_2 NH_3					Partial conversion	
(13)		Zr metal button (10 g) in a water cooled copper crucible. Transfered arc on the crucible.	N_2 1.5–4	60–110	55–75	4.7	Powders are collected on the cold walls of the reactor. Products formed in the crucible with molten metal are analysed by Xrays. Diffusion coefficients of N in molten Zr are determined. (comparision with theory) Superconducting temperature of nitrides is determined		[272] [273] [274] (1968) (1972) (1975)

(14)		Zr metal button (10 g) in a water cooled crucible. Transfered arc on the crucible	Ar 2.5 0–4.5	60–120			The concentration distributions in the crucible products are analysed by X ray. Dissolved amount of oxygen linearly increases with reaction time and air flow rate. Dissolved amount of nitrogen increases parabolicaly with reaction and air flow rate. Nitrogen penetrate in the inner region of the metal, but oxygen stay in the surface region.		275) (1972)
(15)		Zr metal cylinder in a water cooled crucible. Non-transfered arc	N_2 4	70	150	10.5	Products are found in the crucible and on the cold walls of the reactor	87% after 20 min with crucible at two cm from nozzle exit	268) (1968)
(16)		ZrO_2 in the jet of an HF plasma	Ar—N_2 (95% N_2)						276) (1964)
TaN (17)		A Ta metal rod in a water cooled crucible Transfered arc.	N_2 1–4	20–120	80	1.9 to 9.6	Powders are condensed on a water-cooled coil placed arround the reaction zone	Mainly TaN, some Ta subnitrides, Ta	277) (1966)
(18)		Ta chloride + N as carrier gas	H 74	200	117	23 Energy yield 45%	Fine divided powder in the water-cooled reactor		278) (1966)
AlN (19)		Al metal rod in a water cooled crucible. Non-transfered arc	N_2 2.8	50	150	7.5	Powders are concensed on the cold walls of the reactor	98% after 4–7 mn at one cm from nozzle exit	268) (1970)
(20)	From Al	A cube of metallic Al ($10 \times 10 \times 10$ mm^3) in a water-cooled crucible	N_2 1.5–4	60–100	60	3.6–6	A dendrite crystal grew from the metal surface. Powders are collected on the cold walls.	Crucible products 37,6% Powders 5.86%	279) (1968)

Table 15. (continued)

Nitride (case)	Reactants, Carrier gas	Plasma gas (l/min)	I (A)	V_a (V)	P (kW)	Quenching techniques Form of final products	Reaction yield	Ref. (year)
(21)	Compact Al powder in a water-cooled copper crucible Non-transfered arc	N_2 25	250	70	17.5	Powders are collected on the cold walls. A dendrite crystal grew from the molten metal in the crucible.	98% in the crucible 50% for powders (Al as major impurity)	[6] (1976)
(22)	Al_2O_3—C(s) pellet in a water cooled copper crucible. Transfered arc.	N_2 1.5–4	60–100		3–6	Powders are collected on a water-cooled copper coil placed around the reaction zone and in the crucible	70% AlN in the crucible with unreacted Al_2O_3. 16% AlN on the coil	[280] (1968)
(23) From Al_2O_3	Al_2O_3—SiO_2C pellet in a graphite crucible. Transfered arc.	N_2 1.5–4	50–100			Powders are collected on a water-cooled copper coil placed around the reaction zone, and on the crucible.	Al_2O_3—AlN—SiC on the anode. SiO (+Si, SiC as impurities) on the coil	[281] (1971)
(24)	Al powders + $AlCl_3$ carried in NH_3	N_2				Corrosion problems		[282] (1968)
Si_3N_4 (25)	Si cylinder in a water-cooled copper crucible. Non-transfered arc	N_2 2.8	90			Powders are collected on the cold walls of the reactor.	55% α—Si_3N_4 mainly) after 10 min at three cm from nozzle exit	[268] (1968)
(26)	Si powder compacted in a water-cooled copper crucible. Non-transfered arc	25	250	70	17.5	Powders collected on the cold walls. Almost unreacted silicon in the crucible.	100% α—Si_3N_4 on the cold walls (analysis after annealing in Ar)	[6] (1976)

Si_3N_4 (27)	Si powders (75 μ) injected in the plasma jet	N_2 33		17	Powders are collected on the cold walls and in the gases passing out of the plasma zone	Up to 54.11% Si_3N_4 (chemical analysis). X-rays analysis shows that amorphous nitride is formed.	[283] [284] (1978) (1977)
(28)	Si powders (15–35 μm) (15 g/h) injected in an NH_3—HF plasma	NH_3 10.5	HF Plasma 8.2 MHz	10	Powder are collected on a water cooled stainless steel cylinder down-stream from the jet.	Very fine powders are collected. 85% Si_3N_4 (amorphous) The authors have also synthesized AlN and SiAlON with a very good yield.	[285] (1976)
(29)	3 $SiCl_4$ + 16 NH_3 injected down stream the arc 7 mole h^{-1} ($SiCl_4$)	H_2 100		30	Si_3N_4 powder is collected in the water cooled reaction chamber together with NH_4Cl powders. They are separated by sublimation. Si_3N_4 powders are amorphous. They can be partially crystallized to α—Si_3N_4 by heating for one hour at 1150 °C in vacuum.		[289] (1974)
W_2N (30)	W powders injected down stream the arc (500 mesh)	N_2 5		6	Cold quenching tube (1/4" o.d.) below the powder feed ring	25% conversion to W_2N	[261] (1960)
(31)	W metal rod placed in a water cooled copper crucible. Transfered arc.	N_2 1.5–4		6–9	Powders are collected on the water-cooled copper coil placed around the reaction zone.	34% of W_2N in the collected powders	[286] (1968)
Hg_3N_2 (32)	Hg powders injected down stream the arc (2 g/min)	N_2 2.5		12–15	Powders are simply collected in the water-cooled reaction Vessel	40% conversion to Hg_3N_2	[261] (1960)

Table 15. (continued)

Nitride (case)	Reactants, Carrier gas	Plasma gas (l/min)	I (A)	V_a (V)	P (kW)	Quenching techniques Form of final products	Reaction yield	Ref. (year)
Mo_2N (33)	Mo metal rod placed in a water cooled copper crucible. Transfered arc.	N_2 1.5–4			5–8	Powder are collected on the water-cooled copper coil placed around the reaction zone.	40% in the collected powders	[286] (1968)
HfN (34)	Hf metal rod placed in a water cooled copper crucible. Non transfered arc.	N_2 4.6	90			Powders are collected on the cold walls of the reactor	81% conversion after 30 min of reaction at 1–2 cm from nozzle exit	[268] (1968)
BN (35)	Boron powders injected in the jet	NH_3 25	110	135	15			[287] (1962)

→ AlN(g) is lightly favoured and we have consequently a low AlN(g) molar fraction. Since aluminium has quite a low ionization potential, the ionization of aluminium begins at low temperatures (this is an important point for spectroscopic diagnostics).

The situation is completely different for silicon: silicon nitride is unstable at a temperature lower than the boiling point of pure silicon and consequently Si_3N_4 decomposes to give nitrogen and liquid silicon at high temperatures. In other words, the condensation of silicon nitride vapours (SiN, Si_2N) is not favourable to the formation of solid Si_3N_4.

5.2.2 Experimental Results

The experiments summarized in Table 15 can be divided into three groups according to the physical nature of the reactants:

1st group: methods where the reactants are powder fed into the jet
2nd group: methods where the reactants are placed in a crucible
3rd group: methods using gaseous chlorides (more generally a halide)

Before dealing with any particular experiment, we would like to point out the fact that the analysis of final products is a problem that must be carefully studied. In most cases this is the only quantitative information available concerning the synthesis, and conclusions have to be reached from these measurements. Taking into account the high quenching rates in a plasma jet, non-stoichiometric or amorphous products are frequently encountered, and these are difficult to analyse.

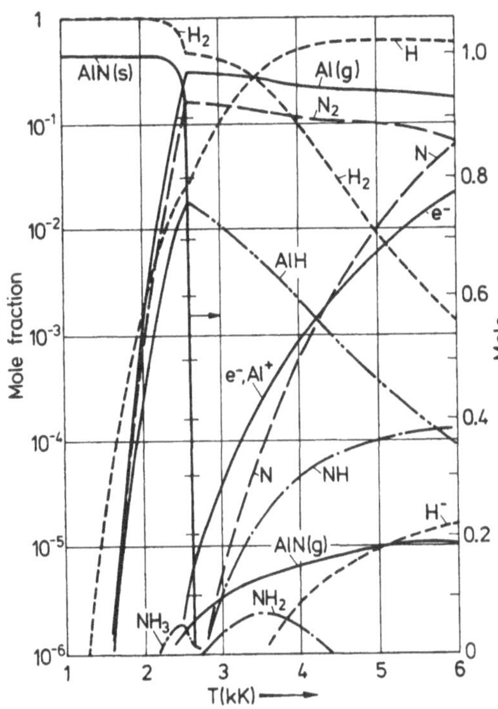

Fig. 65. Equilibrium composition of the mixture (NH_3—Al) versus temperature at 1 atm. Reprinted from [6]

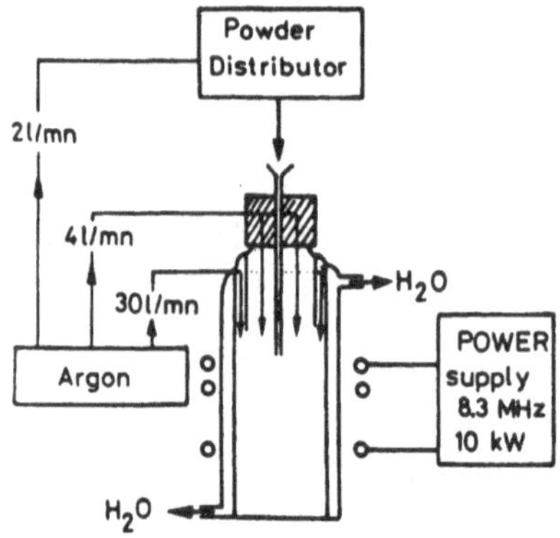

Fig. 66. Experimental set up for the Si_3N_4 synthesis in a N_2 RF plasma. Reprinted from [285]

5.2.2.1 Nitride Formation from Powders

The main difficulty in this kind of experiment lies in the heat transfer from plasma to the solid reactant. If one wants to melt the particles, these have to be sufficiently fine and fed at a low rate, at the right point with a proper velocity. For example Canteloup and Mocelin [285] have carefully chosen the particle size and plasma gas for their experiments on Si_3N_4 synthesis. (Table 15, case 28). They indicated that only 10–15% of the powder was nitrided working with nitrogen in contrast to a conversion of 85 wt · % when working with ammonia as plasma gas. They attributed this difference to the low concentration of monoatomic nitrogen in the N_2 plasma. We do not agree with this interpretation based on the equilibrium diagram of NH_3—Al (Fig. 65) which is similar to the one for Si—NH_3 and not so different from the diagram (Fig. 63) with respect to aluminium nitride. It is likely that the important difference is due to the thermal conductivity of ammonia which enhances the heat transfer to the particles.

The results obtained by Canteloup and Mocelin [285] and by Szymanski et al. [283, 284] on Si_3N_4 synthesis are interesting with respect to the phase obtained in the final products. Silicon nitride (cases 25 to 29) has seldom been obtained as the β phase, generally an amorphous product is formed, indicating reactions through gas phase.

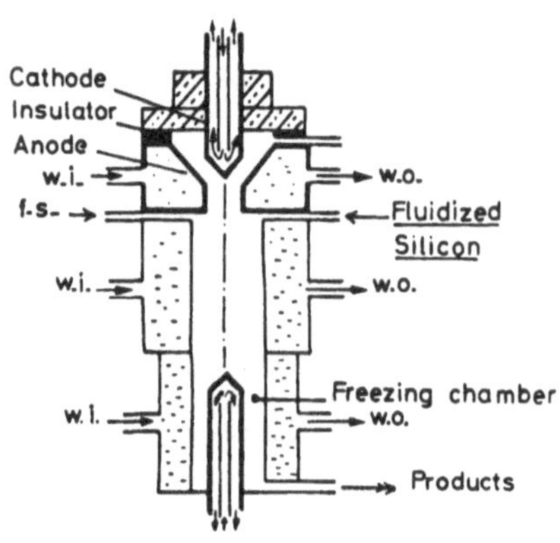

Fig. 67. Experimental set up for the Si_3N_4 synthesis in a N_2 DC plasma jet. Reprinted from [283]

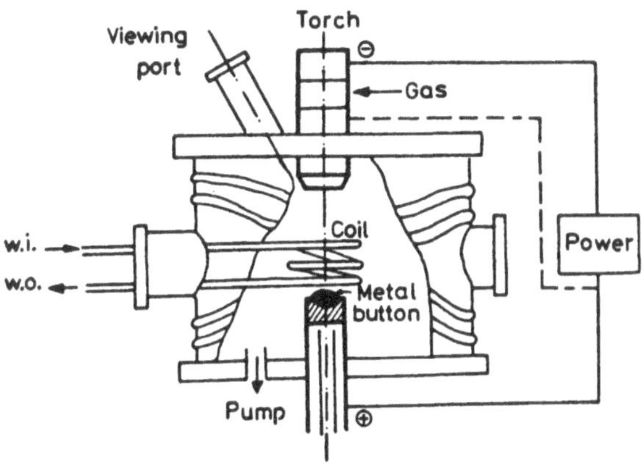

Fig. 68. Experimental set up for ceramics synthesis in a crucible heated by a DC plasma jet with the arc transferred on the crucible. Reprinted from [290]

Figures 66 and 67 show the experimental set up for the two investigations cited. They are now classical and almost every experiment of this type in an arc or HF plasma would be of the same kind.

5.2.2.2 *Nitride Production in a Crucible*

The use of a crucible instead of feeding the powder in the arc has been extensively investigated by Matsumoto, with a transfered arc on the crucible as shown in Fig. 68. The advantage of this approach is that the heat transfer problems are avoided and

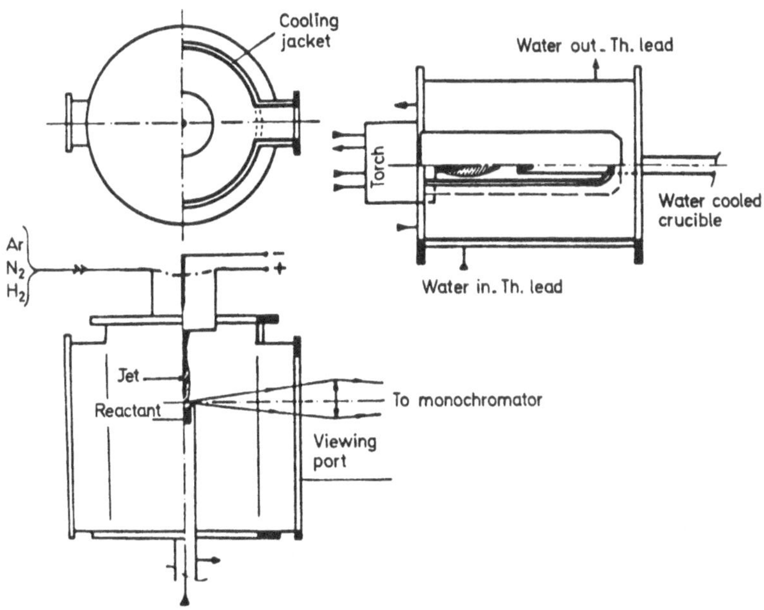

Fig. 69. Experimental set up for ceramics synthesis in a crucible heated by a DC plasma jet. Reprinted from [6]

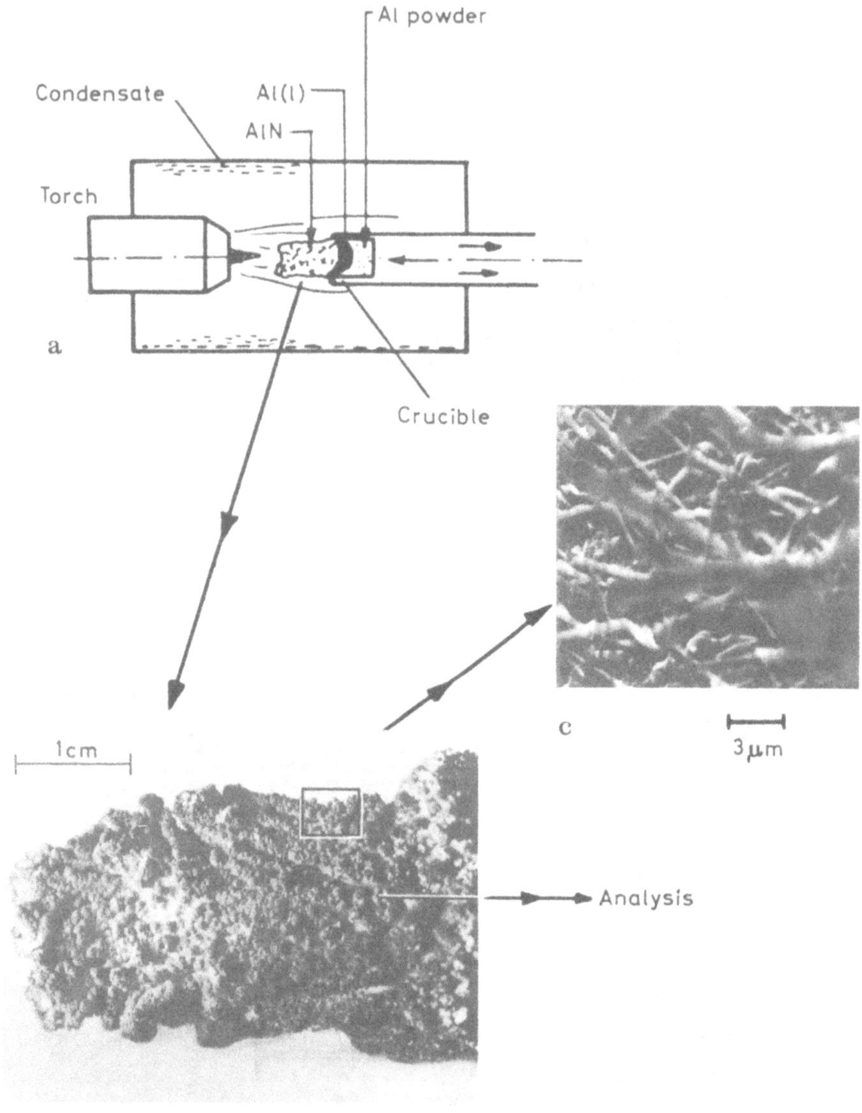

Fig. 70. AlN synthesis by heating Al in a crucible with a DC nitrogen plasma jet (**a**), a photograph of the product (**b**) and a photomicrograph of the product AlN (**c**). Reprinted from [6]

the reaction proceeds in a highly ionized medium at a temperature higher than that in free jets. Matsumoto has analysed his results in two different ways depending upon the location of the products after the reaction. He has analysed the powders condensed on the cooled coil located around the reaction zone and these results are comparable to those of other authors. However, different results are obtained in the case of the dissolution of nitrogen in the liquid metal anode, and these will be considered separately.

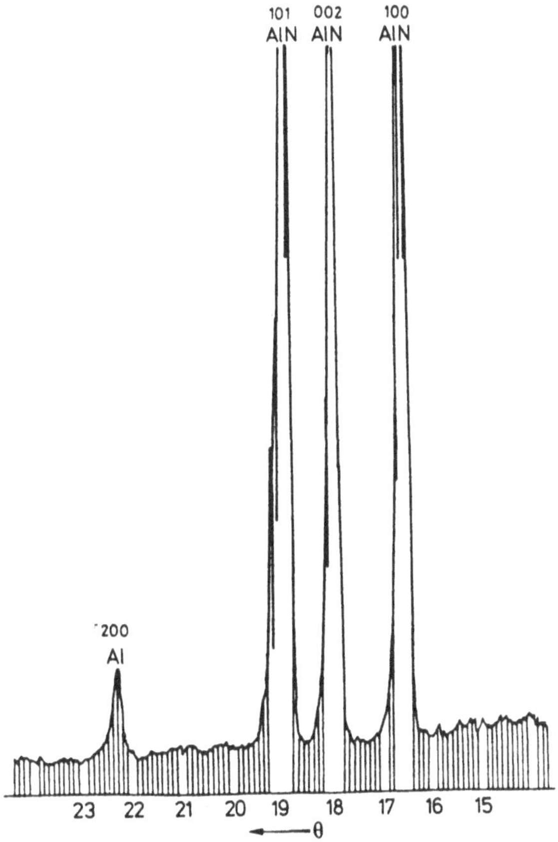

Fig. 71. X Ray diffractogram of the AlN obtained in the crucible. Reprinted from [6] (CuK_α-radiation)

We have done some experiments with a crucible (Fig. 69 and 70a) in a free nitrogen plasma jet [6] using an arrangment by Hayashi et al. [267]. We have preferred to use powders instead of bulk solid in order to have a higher reaction area. An interesting observation during the synthesis of aluminium nitride is the growth of dentrites out of the crucible (Fig. 70b) from liquid aluminium. This phenomenon, also observed by Matsumoto [279] is in agreement with thermodynamical considerations. This dendritic AlN is almost pure nitride as shown by the X diffractogram in Fig. 71. The condensed vapours on the cold walls contained considerable unreacted aluminium together with AlN as predicted by theoretical considerations (Fig. 72).

When silicon is used in our crucible (Fig. 73a) the dentrites are not formed because liquid silicon in the crucible is at a temperature higher than the dissociation point of Si_3N_4. Beta silicon nitride is found with unreacted silicon near the cooled edges of the crucible (a diffractogram of these products is shown in Fig. 73b), however the very fine condensed products on the cold walls are, as previously stated, amorphous. The diffractogram of the resulting fibrous product after heating for one hour at 1200 °C in argon, (Fig. 74) shows the amorphous phase has crystallised to $\alpha-Si_3N_4$. The infrared absorption spectrum of these products, shown in Fig. 75, is in agreement with that obtained by Szymanski ut al. [283] for $\alpha-Si_3N_4$.

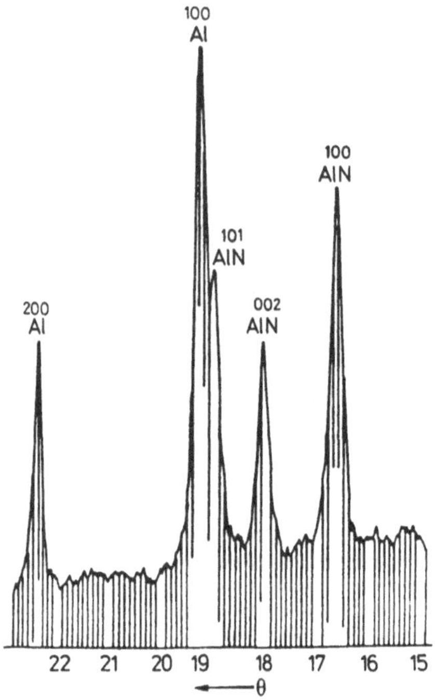

Fig. 72. X-Ray diffractogram of the AlN obtained by condensation of vapors on cold walls. Reprinted from [6] (CuK_α-radiation)

During the occurse of our experiments, we have recorded the emission spectrum of the vapours in the reaction zone. Our spectra have not been calibrated and consequently are only of qualitative interest. Figure 76 shows the spectrum of AlN vapours, and Fig. 77 shows AlO emission bands together with AlN. This latter spectrum was recorded during the reaction proceeding in air. This work, while still not complete, reveals an interesting possibility to analyse the products during the course of a plasma ceramics synthesis reaction.

5.2.2.3 Nitride Synthesis from Chlorides

The third group of techniques consists of reacting a chloride with the nitrogen-hydrogen or ammonia plasma, as also used in classical synthesis. It has been originally tested in plasma torches dealing only with gas phase reactions, followed by condensation. We believe that this is probably the most reliable method for spectroscopic diagnostics of gas phase reactions. Exell et al. [289] have obtained very fine amorphous Si_3N_4 powders using the experimental arrangement shown in Fig. 78. We have computed the high temperature equilibrium of $SiCl_4$—N_2 and of $SiCl_4$—N_2—2 H_2 shown in Figure 79 and 80, respectively. On these diagrams, one can see that SiN(g) has hardly a molar fraction greater than 10^{-3} in both cases. This shows that hydrogen does not sufficiently reduce the chloride to inprove the formation of silicon nitride, and that the wt · % conversion is low, resulting in a considerable by-product formation (NH_4Cl) as shown by the results of Exell et al. [289].

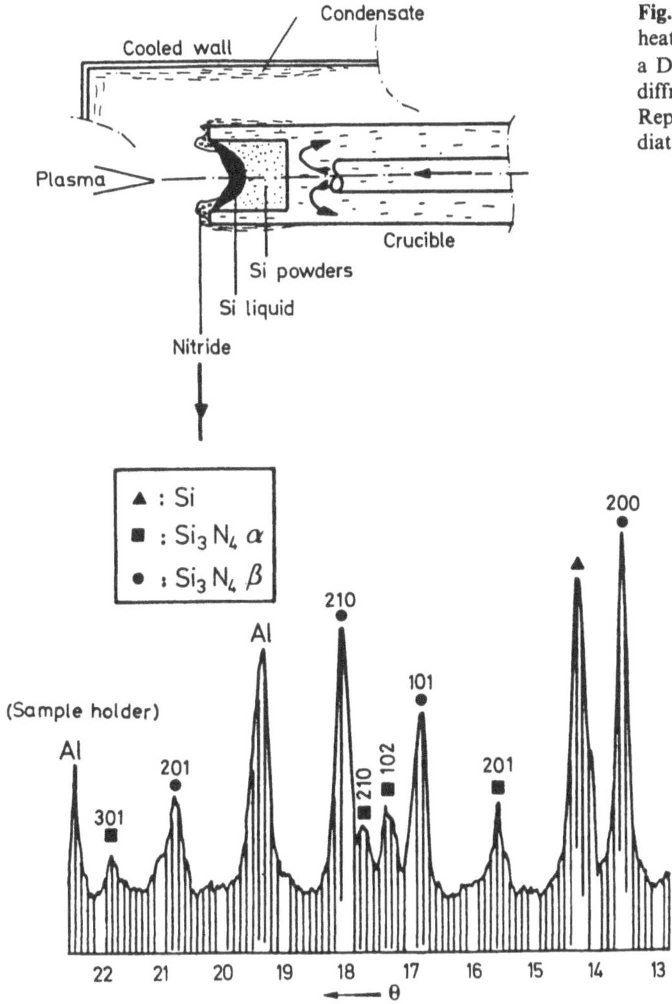

Fig. 73. Si_3N_4 synthesis by heating Si in a crucible with a DC nitrogen jet and X-ray diffractogram of the product. Reprinted from [6] (CuK_α-radiation)

5.3 Carbide Synthesis

Carbides are used as abrasive or as protective materials in nuclear technology and, as for nitrides, plasma technology is mainly used to obtain very fine powders. Experimentally nitride and carbide synthesis in plasmas seem very similar; looking at the references one can notice that plasma chemists have frequently worked with the two materials. Many plasma carbide processes are now patented owing to their important industrial interest and we have summarized in table form (Table 16) the main experimental results on thermal plasma carbide syntheses.

One can notice that the reactions are generally completed and that researchers have been mostly concerned with the structure of the plasma-prepared products. As for nitrides, it is possible to divide the experimental techniques into three groups.

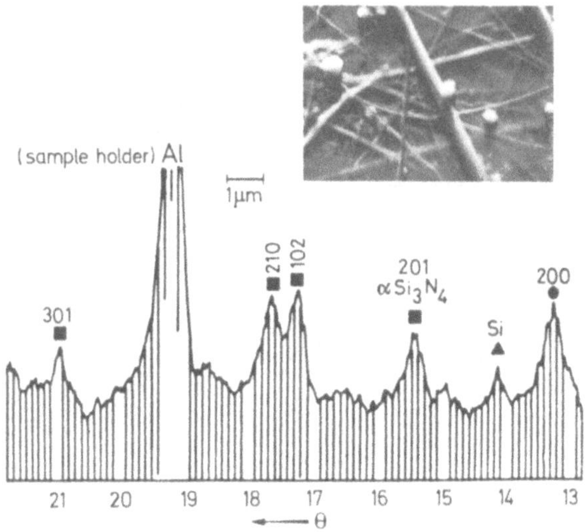

Fig. 74. Analysis of the Si_3N_4 powders obtained by condensation of vapors on cold walls. Reprinted from [6] (CuK_α-radiation)

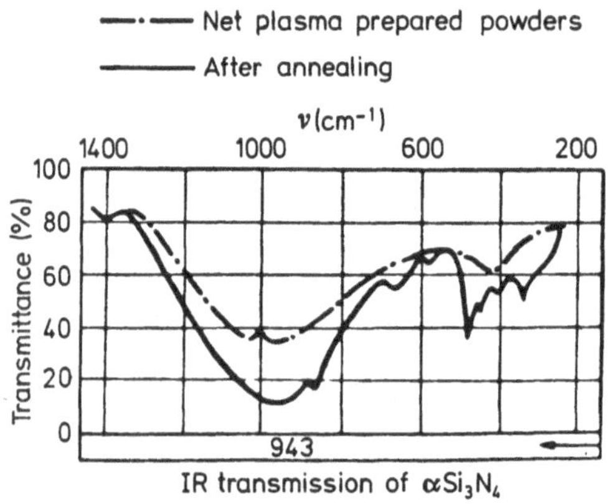

Fig. 75. IR analysis of the Si_3N_4 powders obtained by condensation of vapors on cold walls. Reprinted from [6]

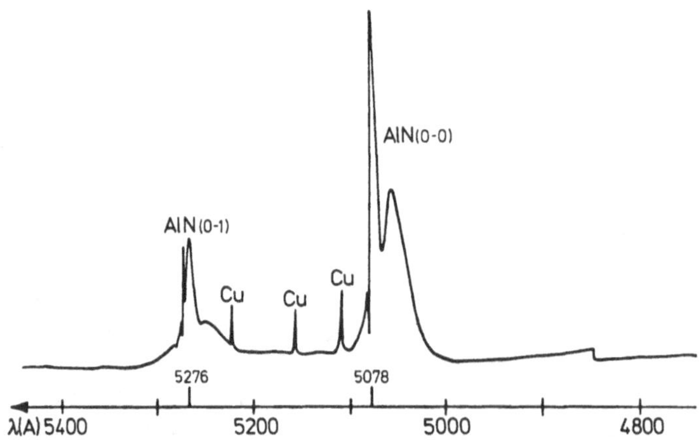

Fig. 76. AlN emission spectrum during the reaction of a nitrogen plasma jet with aluminium powders placed in a crucible. The reaction is performed in a controlled atmosphere. Reprinted from [6]

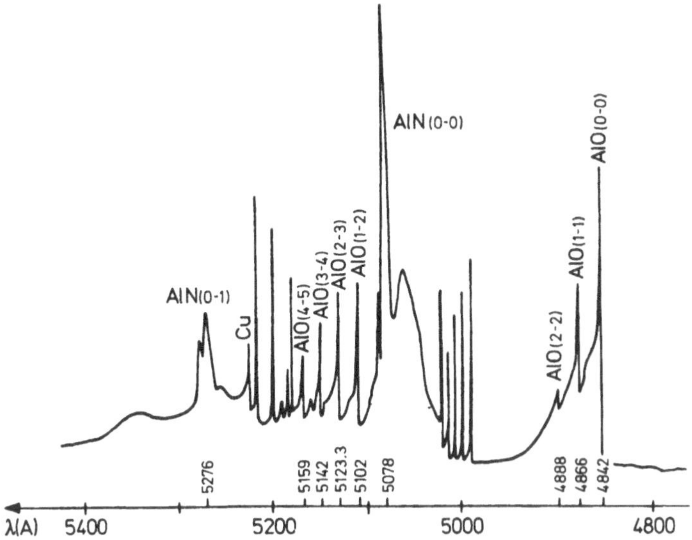

Fig. 77. AlN and AlO emission spectra during the reaction of a nitrogen plasma jet with aluminium powder placed in a crucible. The reaction is performed in ambient atmosphere. Reprinted from [6]

5.3.1 Carbide Synthesis from Powders

This technique has been used by Stokes and co-workers [3] among others for tantalum and tungsten carbide synthesis and the authors have been concerned with a detailed analysis of the influence of the quenching rate on the form of the final product. This work has been extensively reported in Ref. [3]. Perhaps it would now be possible to analyse the experimental data taking into account thermodynamic and transport considerations.

O. Matsumoto [298] has used the powder technique to obtain the cubic phase of molybdenum and tungsten carbides. The feed powder, carried in argon, was hexagonal molybdenum or tungsten carbides formed in plasma with a mixture of MoO_3

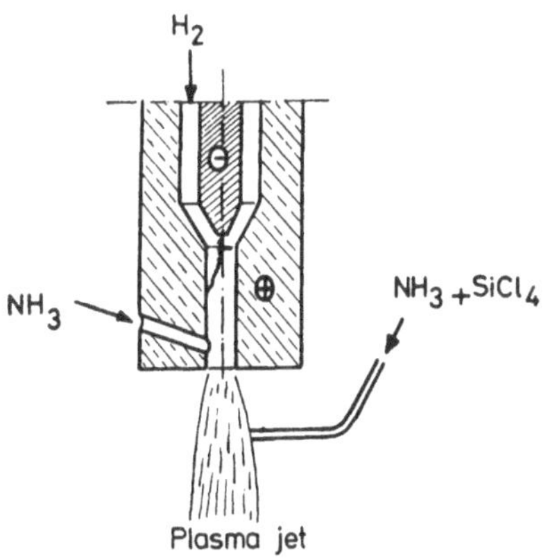

Fig. 78. Experimental set up for the synthesis of carbides and nitrides from chlorides in a DC hydrogen plasma jet. Reprinted from [289]

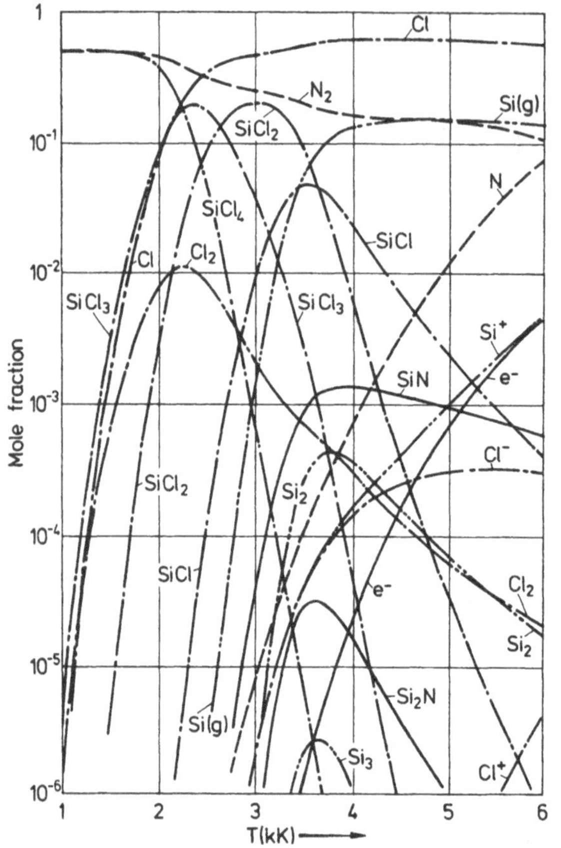

Fig. 79. Equilibrium composition versus temperature for $SiCl_4$—N_2 at one atm. Reprinted from [293)]

and C or WO_2 and C using other methods (see case 17 in Table 16). The cubic phases occur at 2200 °C and 2500 °C for αM_0C_{1-x} and βWC_{1+x} respectively, and necessitate a rapid quenching rate to be obtained at ordinary temperature.

5.3.2 Synthesis of Carbides from Solids in a Crucible

This kind of technique has been mainly used by Matsumoto and by Bosch and De Vynk [264)]. A mixture of metal and carbon or of metal oxide and carbon is heated in a plasma jet of varying composition and the effect of the mixture ratio is studied together with the reaction time. The percent conversion is generally good but fine powders are not formed by this way.

An interesting method similar to Matsumoto's has been proposed by Wickens [301)] for boron carbide synthesis. The experimental arrangement is shown Fig. 81; a high intensity arc is struck between a graphite cathode and a composite consumable anode made of boron oxide or sodium borate which are more readily available and cheaper than boron trichloride. B_4C was obtained by this method, but the author indicates that the power comsumption (kWh/kg B_4C) was high owing to insufficient erosion of the consumable anode.

Table 16. Carbides Synthesis

Carbides (case)	Reactants, Carrier gas	Plasma gas (l/min)	I (A)	V_a (V)	P (kW)	Quenching techniques Form of final products	Reaction yield	Ref. (year)
ZrC								
(1)	(ZrO_2 + C) Solid mixture in the plasma jet	Ar	HF Plasma			ZrC has been obtained depending on the ZrO_2/C mixture ratio		[276] (1964)
(2)	ZrO_2 powders fed in the jet	Ar + C_6H_6				ZrC has been obtained		
(3)	$ZrSiO_4$ + C Mixtures in a water cooled crucible transfered arc	4 Ar Ar + H_2	150			The products are found in the crucible in solid form and on the cooled coil as powders	Up to 10.8 wt% of the crucible products arc ZrC. Other products are Zr, Si, SiC, SiO_2, C. Powders on the coil are SiO	[291] (1973)
(4)	ZrH_2 + C or ZrO_2 + C solid mixtures in a crucible. AC arc between anode and crucible	He 15	300 DC + 200 AC super imposed	35			ZrC with unreacted ZrO_2, no Zn metal detected	[264] (1970) (1971)
TaC								
(5)	Ta_2O_5, C solid mixtures in a water-cooled Cu crucible. Transfered arc	Ar				The analysed products are found in solid form in the crucible	The identified products are TaC, Ta_2C depending on the C/Ta_2O_5 mixture ratio	[292] (1975)
(6)	Tantalum chloride fed together with CH_4 in H_2 plasma	H_2 74	200	117 V	10.5 kW (45%)	Products are quenched in a water-cooled vessel		[278] (1966) [293] (1967)

Table 16. (continued)

Carbides (case)	Reactants, Carrier gas	Plasma gas (l/min)	I (A)	V_a (V)	P (kW)	Quenching techniques Form of final products	Reaction yield	Ref. (year)
(7)	Ta or Ta_2O_5 powders feeded which CH_4 in an helium plasma						For the best experimental conditions 72% TaC, 14% Ta_2C conv. from Ta 24% Ta, 17% TaC conv. from Ta_2O_5	[3] (1971)
TiC (8)	$TiCl_4$ carried by $H_2 + CH_4$ or C_2H_2 carried by H_2	20	80	90	7.2	Products are quenched in a quenching tube. Powders 0.02–0.35 µm	TiC with a yield of 98.2% of the theoretical yield	[294] (1970)
(9)	$TiCl_4 + H_2$ and CH_4 6 g/min $TiCl_4$ 1 g/min CH_4	H_2					92% TiC	[293] (1967)
(10)	$TiCl_4 + H_2 + CH_4$ 0.12 mole/min $TiCl_4$ + 10 l/min H_2 8.5 l/min CH_4	H_2 100			30	Products are quenched in a water-cooled vessel 100 Å- to 400 Å powders	70% conversion to TiC	[289] (1974)
(11)	TiH_2 + C or TiO_2 + C in a crucible. AC arc between anode and crucible	He 15	300 DC + 200 AC	35 80			9 kW/kg DC + + 13 kWh/kg AC TiH_2 is completely converted into TiC. TiO_2 is completely converted into TiC with carbon impurities	[264] (1970 to 1971)

SiC (12)	$SiCl_4 + CH_4$ in an H_2 plasma $CH_4/SiCl_4$, mole. ratio of 5	H_2				Powders in the water-cooled vessel 100–200 Å powders	Non-pyrophoric SiC powders	[289] (1974)
(13)	$SiCl_4$ or $SiCl_3CH_3$ or $SiCl_2(CH_3)_2$	$H_2 + A_r$ vol.% (80, 20) 50			19	Powders in the water-cooled vessel 500 Å powders	Almost pure β-SiC (with 1% α-SiC)	[295] (1977)
(14)	Methylchlorosilanes + Ar fed down stream from the jet	Ar, H_2 121 scFh 44 scFh	HF Plasma 50 kW, 4 MHz			Products are collected in a filter placed down stream from the jet; Powders of 1 µm approx.	β-SiC 80% of the total products	[296] (1972)
(15)	Si + C or SiO_2 + C in a crucible. AC arc between anode and crucible	He 15	300 DC 200 AC	35 80		Products contain unreacted Si when working with (Si, C) and unreacted SiO_2 when working with (SiO_2, C)	β-Si 14,6 kWh/kg DC + 20,8 kWh/kg AC β-Si 17,5 kWh/kg DC + 25 kWh/kg AC	[264] (1970 to 1971)
NbC	Nb_2O_5/C solid mixture in a water-cooled crucible. Transfered arc.	Ar 4	130			A silvery-white metallic cake is obtained on the graphite hearth anode	NbC, Nb_2C 10 wt% unreacted Nb graphite 1 wt%	[297] (1974)
MoC (17)	C/MoO_3 solid mixtures and C/(MoO_3 + TiO_2) solid mixtures in a WC crucible	Ar 4				A silvery-white metallic cake is obtained on the graphite hearth anode	Hexagonal or a mixture of Hexagonal and cubic MoC is obtained with MoO_3/+ C. Cubic $Mo_xTi_yC_z$ is obtained with (MoO_3 + + TiO_2) + C	[299] (1974)

Table 16. (continued)

Carbides (case)	Reactants, Carrier gas	Plasma gas (l/min)	I (A)	V_a (V)	P (kW)	Quenching techniques Form of final products	Reaction yield	Ref. (year)
(18)	Mo_xC_y plasma prepared powders (see preceding case) feeded in Ar jet	Ar 20 (.5 g/min of powders)			2.2	The powders are quenched in a water-cooled reactor	(α)-MoC_{1-x} verconducting temperature are studied	298) (1976) 300) (1977)
(19)	No/C mixture or NoO_3/C mixture in a crucible AC arc between anode and crucible	He 15	300 DC 200 AC	35 80		Mo_2C with unreacted Mo in the case of Mo + C Mo_2C, MoC, Mo and unreacted MoO_3 in the case of MoO_3 + C	7 kWh/kg DC 8 kWh/kg AC	264) (1970)
WC								
(20)	WC—W_2C plasma prepared powders (see case 18) fed in Ar jet	Ar 20			2.2	The powders are deposited in a water-cooled reactor	β-WC_{1-x} was obtained with traces of WC and α-W_2C	298) (1976)
(21)	W/C solid mixtures or WO_3/O mixtures in a drucible. AC arc between anode and crucible	He 15	300 DC 200 AC	35 80			α-W_2C and WC were formed with traces of unreacted W and C (see case 19)	264) (1974)
(22)	W or WO_3 powders fed together with methane.	He					10% conv. to WC 40% conv. to W_2C starting with W powders. 10% WC 30% W_2C starting with WO_3 powders + 40% W unreacted	3) (1971)

B_4C	(B_2O_3, C) or ($Na_2B_4O_7$, C) in the consumable anode of a high intensity arc	Ar + H_2 25 12	200 300	65	12 20	The products are quenched on a water cooled probe	The deposited products are B_4C with C	[301] (1976)
(24)	$BCl_3/H_2/CH_4$ mixtures fed in an argon plasma jet	Ar	HF Plasma 30 kW	5 MHz	28	Products are quenched in a water cooled tube	The highest, conversion of BCl_3 to boron carbide was obtained at low BCl_3 flow rate 20 g/min, high H_2/BCl ratio of 8, stoichiometric BCl_3/CH_4 ratio of 4.	[302] (1975)
						The authors have extensively studied the reaction conditions and have suggested a reaction mechanism for boron or boron carbide formation from BCl_3.		

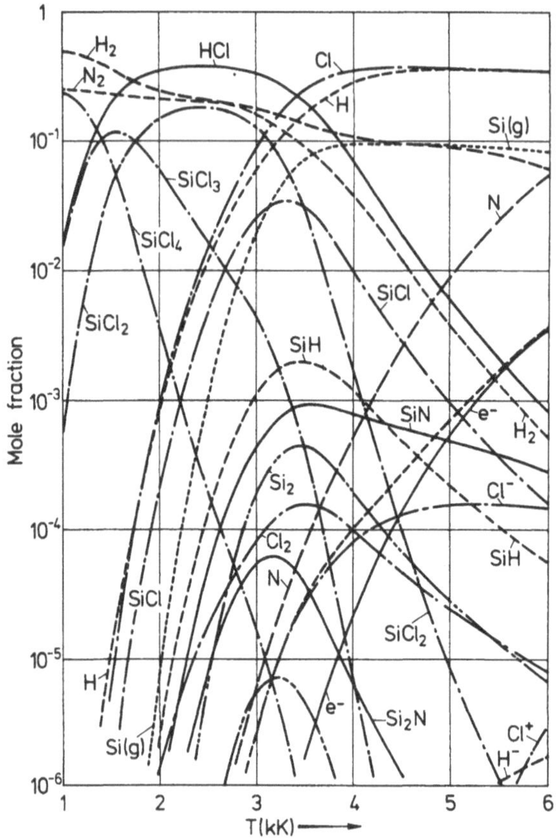

Fig. 80. Equilibrium composition versus temperature for $SiCl_4$—N_2—H_2 at one atm. Reprinted from [293]

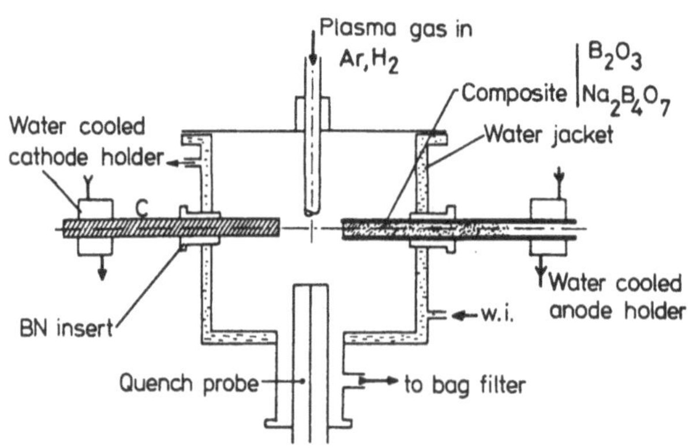

Fig. 81. Experimental set up for boron carbide synthesis with a high intensity arc between a graphite cathode and a consumable anode with boron oxide. Reprinted from [301]

5.3.3 Synthesis of Carbide from Chlorides

This method is mostly used to achieve a good yield of fine powders. A chloride is fed together with methane (or other hydrocarbon), in the tail flame of an Ar, or He plasma jet. (Fig. 79). Neuenschwander patented this method in 1967 [293].

To reduce the power consumption Salinger [296] and O. de Pous [295] have proposed the use of methylchlorosilane instead of the chloride. MacKinnon, Reuben and Hamblyn [302, 303] have reported a very interesting work on boron and boron carbide synthesis from BCl_3 in an HF plasma. Using thermodynamics and condensation kinetics calculations they propose a reaction mechanism which excludes the vapour phase formation of boron carbide. They suggest that carbon nuclei form first on a cold surface and then react with BCl molecules to give boron carbide. Perhaps the same considerations could be applied to other sytems to improve their performance.

Other studies concerning the chloride route worth mentioning here are for $TiCl_4$ [304, 305] and BCl_3 [305, 307].

6 References

1. Venugopalan, M.: Plasma Chemistry. An Introduction. In: Reactions under plasma conditions. Ch XI. Vol. II (ed. Venugopalan), Wiley Interscience, New York 1971
2. Kingery, W. D., Bowen, H. K., Uhlmann, D. R.: Introduction to ceramics, Wiley Interscience, New York 1975
3. Stokes, C. S.: Chemical Reactions in Plasma Jets, in: Reactions under plasma conditions, Ch XV. vol II (ed. Venugopalan), Wiley Interscience, New York 1971
4. Zeleznick, F. J., Gordon, S.: Ind. Engng. Chem. *60*, 6, (1968)
5. Storey, S. H., Van Zegeren, F.: The computation of chemical equilibria, Cambridge University Press 1970
6. Bourdin, E.: Contribution à l'étude théorique et expérimentals de la syntheèse de nitrures par réaction d'un jet de plasma d'azote avec des poudres d'aluminium et de silicium. Thèse de 3ème cycle, Univers. Limoges, 1976
7. Fauchais, P., Baronnet, J. M., Bayard, S.: Rev. Int. Htes. Temp. et Réfract. *12*, 221, (1975)
8. Bourdin, E., Aubreton, J., Fauchais, P.: Equilibrium composition and thermodynamic properties of $C-H_2$, CH_4 and CH_4-H_2 plasmas, computation of their reducing action on iron oxyde Fe_2O_3. In: IIIrd Internat. Symp. plasma Chemistry, Limoges july 1977, Proc. III-G4-2
9. Drellishak, K. S.: Partition functions and thermodynamic properties of high temperature gases, Ph. D. Northwestern Univ. 1963
10. Coldwell, D. M.: High Temperature Science *8*, 309, (1976)
11. Mason, E. A., Vanderslice, J. T., Yos, J. M.: Phys. of Fluids *2*, 688 (1959)
12. Hirschefelder, J. O., Curtiss, C. F., Bird, R. B.: Molecular Theory of Gases and Liquids, J. Wiley and sons, Inc., New York 1964
13. Devoto, R. S.: Phys. of Fluids *10*, 2105 (1967)
14. Mason, E. A.: Kinetic processes in gases and plasmas (ed. A. R. Hochstim), Academic Press, N.Y. and London, 1969
15. Dalgarno, A.: Phil. Trans. Roy. Soc. London *A250*, 426 (1958)
16. Chapman, S., Cowling, T. G.: The mathematical theory of non uniform gases, Cambridge Press, U.K. 1970
17. Athye, W. F.: A critical evaluation of methods for calculating transport coefficients of partially and fully ionized gases, NASA TN, ND-2611, 1965
18. Devoto, R. S.: Phys. of Fluids *9*, 1230 (1966)
19. Devoto, R. S., Li, C. P.: J. Plasma Physics *2*, 17 (1968)
20. Devoto, R. S.: A.I.A.A. J. *7*, 2 (1969)
21. Devoto, R. S.: Phys. of Fluids *16*, 616 (1973)
22. Gorse, C.: Contribution au calcul des propriétés de transport des plasmas des mélanges argon-hydrogène et argon-azote. Thèse de 3ème cycle, Univers. Limoges, 9 sept. 1975
23. Butler, J. N., Brokow, R. S.: J. Chem. Phys. *26*, 1636 (1957)
24. Vanderslice, J. T. et al.: Phys. of Fluids *5*, 155 (1962)

25. Emmons, H. W.: Modern Developments in heat transfert (ed. W. Ibel), Academic Press New-York (1963)
26. Freeman, M. P.: Chemical Research in streaming thermal plasma, in Adv. High Temperature Chem. (L. Eyring ed.) vol. 2 pp. 151–202 Academic Press, New-York 1969
27. Frank-Kamenestkii, D.: Diffusion and heat transfert in chemical kinetics, Plenum Press. New-York—London (1969)
28. Barrière, M., Prud'homme, R.: Equations fondamentales de l'aérothermochimie, Masson et Cie Paris, 1973
29. Prigogine, I., Defay, R.: Thermodynamique chimique, Dunod, 1946
30. Haase, R.: Thermodynamics of irreversible process, Addison Wesley Publ., 1969
31. Landau, L., Lifschitz, E.: Mécanique des fluides, Edditions Mir., Moscou, 1971
32. Brun, E. A., Martinot-Lagarde, A., Mathieu, J.: Mécanique des Fluides, Dunod, Paris 1968
33. De Groot, S. R., Mazur, P.: Non equilibrium thermodynamics, North Holland, 1961
34. Muster, A.: Thermodynamique des processus irréversibles, Presses Universitaires de France, 1966
35. Williams, F. A.: Combustion theory, Addison Wesley, 1965
36. Lightfoot, E. N.: Estimation of heat and mass transfer rates, Lectures in transport phenomena (Ed. Bird R.B. et al.) A.I.Ch.E. continuing Education Ser. (1969), 345 East 47 Street New-York 10017
37. Meyer, H.: Über das Schmelzen von Pulvern in Plasmastrahl. Max-Planck-Institut für Silicatforsch.), Würzburg 1969
38. Pfender, E.: Pure and Appl. Chem. *48*, 199 (1976)
39. Kanzawa, A., Pfender, E.: I.E.E.E. Transactions on plasma Sciences, vol. PS-6 n° 1 march 1978
40. Petrie, T. W.: The effect of ionization on heat transfer to wires immersed in an arc plasma, Ph. D Thesis Univers. Minnesota 1969
41. Finkelnburg, W., Maecker, H.: Elektrische Bögen und Thermisches Plasma, Encyclopedia of Physics (ed. S. Flügge), vol. 22, p. 254, Springer, Berlin, 1956
42. Eckert, G. R. G.: Electrode components of the arc discharge, Ergebn. d. Exakten Naturwiss., vol. 33, p. 1, Springer, Berlin, 1961
43. Kimblin, C. W.: Anode phenomena in vacuum and atmospheric pressure arcs I.E.E.E. Trans. plasma Sci. *PS-2 (4)* 310 (1974)
44. Eckert, E. R. G., Pfender, E.: Advances in plasma heat transfer, in Adv. Heat Transfer, vol. 4, Academic Press, New York (1967)
45. Incropera, F. P., Leppert, G.: Int. J. Heat Transfer. *10*, 1861, (1967)
46. Ozisik, M. N.: Radiative Transfer, John Wiley and Sons. New York, 1973
47. Sieder, E. N., Tate, G. E.: Ind. Engng. Chem. *28*, 1429 (1936)
48. Dittus, F. W., Boelter, L. M. K.: Univers. Calif. Pub. Eng. *2*, 443 (1930)
49. Martinelli, R. C.: Trans. ASME *69*, 947 (1947)
50. Ostrasch, S.: Nat. Adv. Comm. Aeron., Tech. note 2665, Washington D.C., 1952
51. Jakob, M., Dow, W. M.: Trans. ASME *68*, 124 (1946)
52. Huang, G. C.: I. Heat Transfer *85*, 235 (1963)
53. Newton, I.: Philos. Nat. Primo. Math. *2*, Sect. 2, 7 (1710)
54. Stokes, G. C.: Trans. Cambr. Phil. Soc. *9*, Pt 2, 51 (1851)
55. Pruppacher, H. R., Leclair, B. P., Hamielec, A. E.: J. Fluid. Mech. *44*, 781 (1970)
56. Clift, R., Gauvin, W. H.: Motion of Entrained particles in gas streams, Can. J. Chem. Eng. *49*, 439 (1971)
57. Torobin, L. B., Gauvin, W. H.: Can. J. Chem. Engng. *37*, 129, (1959)
58. Boulos, M. A., Gauvin, W. H.: ibid. *52*, 355, (1974)
59. Bhattacharyya, D., Gauvin, W. H.: A.I.Ch.E. J. *21*, 879, (1975)
60. Lewis, J. A., Gauvin, W. H.: ibid. *19*, 982, (1973)
61. Boulos, M. I.: IIIème Symp. Internat. Chimie des Plasmas, IUPAC, S-3-2. Limoges (1977)
62. Borgianni, C. et al.: Combust. Flame *13*, 2, 181, (1969)
63. Bonet, C.: Thermal Plasma Processing, Chem. Eng. Progr. *72*, 12, 63 (1976)
64. Yoshida, T., Akashi, K.: J. Appl. Physics. *48*, 2252 (1977)
65. Fiszdon, J., Lesinski, J.: Internat. Round Table on Study and Application of Transport Phenomena in Thermal Plasmas, Odeillo, 12–16 sept. (1975)

66. Rykalin, R. R.: Pure and Appl. Chem. *48*, 229, (1976)
67. Murray, W. D., Landis, F.: J. Heat Transfer *81*, 106, (1959)
68. Capitelli, M. et al.: Combut. Flame *15*, 23, (1970)
69. Boulos, M. I.: I.E.E.E. T. PS *4*, 28, (1976)
70. Boulos, M. I.: Proc. Int. Round Table on Study and Application of Transport Phenomena in Thermal Plasmas, Odeillo, 12–16 sept. 1975, paper IV-7
71. Munz, R. J.: The decomposition of molybdenum disulphide in an induction plasma tail flame, Ph. D. dissertation, McGill Univ. 1974
72. Vardelle, M.: Contribution à l'étude de la projection de particules d'alumine par plasma: mesure des températures et des vitesses des particules, corrélation avec les structures métastables obtenues, Thèse de 3ème cycle, Limoges Univers., march 1980
73. Warner, D. D.: J. Phys. Chem. *87*, 2329, (1977)
74. Johston, H.: Gas phase reaction rate theory, Ronald Press, New York 1966
75. Gladstone, S., Laidler, J. K., Eyring, H.: The theory of rate processes, McGraw-Hill Book Company, New York and London 1941
76. Emanuel, N., Knorre, D.: Cinétique chimique, Edition Mir., Moscou 1975
77. Polyanyi, J. C.: Acc. Chem. Res. *5*(5), 1961, (1972)
78. Coudert, J. F. et al.: IV Symp. Internat. Chimie des Plasmas, IUPAC Zurich, août 1979, p. 349 (ed. Vepřek)
79. Baulch, D. L., Drysdale, D. D., Morne, D. G., Lloyd, A. C.: Evaluated kinetic data for high temperatures reactions. Vol. II, Butterworths, London 1973
80. Griem, H. R.: Plasma Spectroscopy, Mc. Graw-Hill, NY, 1964
81. Drawin, H. W.: High Pressures, High Temperatures *2*, 359, (1970)
82. Ranson, P., Chapelle, J.: J.Q.S.R.T. *14*, 1 (1974)
83. Ranson, P.: Contribution à l'étude du rayonnement des plasmas d'argon faiblement ionisés: spectres des raies, rayonnement de recombinaison, bremstrahlung et électrons atomes, Thèse de Doctorat d'Etat, Univers. Orléans, 9 febr. 1978 (unpublished — available at CNRS France)
84. Baronnet, J. M.: Contribution à l'étude spectroscopique des plasmas d'azote produits par un générateur à arc soufflé; application à la chimie des plasmas: synthèse des oxydes d'azote, Thèse de Doctorat d'Etat, Univers. Limoges 1978
85. Catherinot, A., Sy, A.: Z. Naturforsch. *30a*, 1143, (1975)
86. Delcroix, J. L.: Physique des Plasmas, Dunod Paris 1963
87. Drawin, H. W.: Reactions under plasma condition (ed. Venugopalan, M.), Wiley Interscience N.Y., 1971
88. Stupochenko, Y. V., Losev, S.. A., Osipov, A. J.: Relaxation in shock waves, Internat. ser. applied physics and engineering, 1, Springer Verlag, 1967
89. Leskov, L. V., Savin, F. A.: Soviet Physics Uspekhi *3*, 912 (1961)
90. Losev, S. A., Osipov, A. L.: ibid. *74*, 525, (1962)
91. Zeldovich, B., Raiser, Y.: Physics of shock waves and high temperature hydrodynamic phenomena, Academic Press N.Y., 1966
92. Rowe, B.: Etude des réactions chimiques et ioniques dans un plasma en fort déséquilibre thermodynamique. Cas du mélange azote-argon, Thèse de Doctorat d'Etat, Univers. Paris VI (1975)
93. Cabannes, F.: Molecular Spectroscopic Diagnostics, in: Analytical Uses of Plasmas (ed. R. M. Barnes), John Wiley N.Y. 1978
94. Bourasseau, D.: Contribution à l'étude de l'équilibre thermodynamique local dans les jets de plasma d'argon sous pression réduite, Thèse de doctorat de 3ème cycle Paris, 1968
95. First report on spectroscopic methods of temperature measurements. IUPAC, Rep. of the committee on standards and measurements in plasma chemistry, (ed. P. Fauchais), Limoges Univers., april 1980
96. Bayard, S.: Contribution au calcul des fonctions de partition des plasmas azote-silicium-aluminium et détermination des températures à partir du fond continu de l'azote, Thèse de 3ème cycle, Univers. Limoges 1974
97. Hofsaess, D.: J.Q.S.R.T. *19*, 339, (1978)
98. Fieffe-Prevot, P.: Rev. Phys. Appl. *12*, 1243 (1977)

99. Olsen, H. N., Bedtai, G., Kelly, F. L.: Development of diagnostic methods for seeded air and nitrogen plasma, A.E.D.C. — T.R. 68-217
100. Ranson, P., Chapelle, J.: J.Q.S.R.T. *14*, 1 (1974)
101. Konjevic, N.: J. Phys. Chem., Ref. Data *5*, 209, (1976)
102. Coudert, J. F.: Contribution à l'étude de la synthèse des oxydes d'azote par chalumeau à plasma, Thèse de 3ème cycle, Université de Limoges (1978)
103. Fiszdon, J., Gauk, , Daniault, : Rev. Int. Htes Temp. et Réfract. *13*, 11 (1976)
104. Hare, A. L.: Velocity measurements in plasma flows using cooled pitot tubes: an unsolved problem, Proc. G. 3.9., IIId Internat. Symp. Plasma Chem., Limoges 1977
105. Cambray, P.: Contribution à la mesure des flux de chaleur et de la vitesse d'écoulement du plasma dans un générateur à arc soufflé, Thèse de 3ème cycle, Poitiers (1971)
106. Guyonnet, J., Borie, A., Fauchais, P.: Projection au chalumeau à plasma de revêtements céramiques à coefficients de frottement variables, in: Proc., IV-5 Internat. Round Table on Study and Application of Transport Phenomena in Thermal Plasmas, IUPAC, Odeillo, 12–16 sept. 1975
107. She, C. Y. et al.: Optics Lett. *2*, 30, (1978)
108. Gouesbet, G.: Phénomènes de diffusion et de thermodiffusion des espèces neutres dans un plasma d'argon et d'hélium. Thèse de Doctorat d'Etat, Univers. Rouen 1977
109. Vardelle, A.: Contribution à la mesure statistique des vitesses et des températures de surface de particules injectées dans un jet de plasma d'arc, Thèse de 3ème cycle, Univers. Limoges 1979
110. Cambray, P.: Mesure de vitesse dans des jets de plasma comportant des gradients axiaux. Proc. G. 3.2., 3d Internat. Symp. Plasma Chemistry, Limoges University 1977
111. Perugini, G.: Analysis of dynamic fusion phenomena of ceramic and metallic powders injected into an argon plasma jet, 3d Internat. Meet. Modern Ceramics Technologies, Rimini, Italy 1976
112. Todorovic, P. S. et al.: Spectrochimica Acta *313*, 103 (1976)
113. André, M.: Les techniques de cinématographie ultra-rapide, C.E.A., Centre de Limeil, France 1977
114. Gold, D.: J. Physics E, Scientific Instruments *10*, 395, (1977)
115. Durst, F., Zare, M.: Bibliography of laser doppler anemometry literature, Univers. Karlsruhe, DISA Information Department
116. Boutier, A.: Data processing in laser anemometry, 3d Internat. workshop on laser velocimetry, Institut Franco-Allemand de Saint-Louis, France 1976
117. Edwards, R. V. et al.: J. Appl. Phys., *42*, 837 (1971)
118. Edwards, R. V. et al.: ibid. *44*, 1694, (1973).
119. Caspersen, C., Saustrap, H., Kristensen: L.D.A. Measurements in a low-velocity wind tunnel on a fin plate model, DISA information *17*, (1975)
120. Self, S. A.: Laser doppler velocimetry in MHD boundary layer, Eng. Aspects Magnet., 14th Symp. Tullahoma Tenn. 1974
121. Steenstrup, F. V.: Disa Information *18*, 21 (1975)
122. Abbiss, J. B.: Seminar at NATO advanced study, institute on photon correlation spectroscopy and velocimetry, Capri (Italy), 26 july/6 aug. 1976
123. Bonet, C.: Description d'une méthode de mesure de la température d'une particule en mouvement, Réunion R.C.P. Gaz Ionisés Réactifs, Paris le 10/02/78 (ed. P. Fauchais) Limoges Univers.
124. Kruszevska, B., Lesinski, J.: Rev. Phys. Appliquée *12*, 1209, (1977)
125. Schnell, C. R. et al.: The industrial application of plasma technology for the production of fused silica, Symp. Commercial Potential for Arc and Plasma Processes, Atlantic City, 8–11 sept. 1974
126. Mackinnon, I. M., Reuben, B. G.: J. Electrochem. Soc. *122*, 806, (1975)
127. Roman, W. C., Zabielski, M. F.: Spectrometric Gas Composition Measurement of UF_6 RF plasmas. 30th Ann. Gaseous Electronics Conf., Palo Alto, CA, oct. 1977
128. Rykalin, N. N.: Pure and Appl. Chem. *48*, 179, (1978)
129. Hamblyn, S. M. L., Reuben, B. G., Thompson, R.: Hydrogen reduction of boron trichloride to boron in an R. F. plasma in special ceramics, (ed. Popper P.), Brit. Ceramic Res. Ass., vol. 5, p. 147

130. Eckert, H. U.: J. Appl. Phys. *48*, 1467 (1977)
131. Gage, R. M.: Arc torches and processes, U.S. Pat. 2, 806, 124 (1957)
132. Bose, T. K., Pfender, E.: A.I.A.A. J. *7*, 1643, (1969)
133. Shih, K. T., Pfender, E.: ibid. *8*, 211, (1970)
134. Cambray, P.: Contribution à la mesure des flux de chaleur et de la vitesse d'écoulement du plasma dans un générateur à arc soufflé, Thèse de 3ème cycle, Univers. Poitiers, 6 april 1971
135. Hassan, H. A., Smith, N. S.: A.I.A.A. J. *8*, 657 (1970)
136. Leontiev, A. J., Voltchkov, E. P.: Caractéristiques électriques et thermiques d'un plasmatron de haute enthalpie, in Investigations expérimentales des plasmatrons (M. F. Joukov, ed.), Nauka, Novisibirsk (1977)
137. Kimblin, S. U.: Erosion des électrodes et processus d'ionisation entre les électrodes de l'arc dans le vide et à la pression atmosphérique, p. 226 in: Investigation expérimentales des plasmatrons (ME Joukov ed.) Nauka, Novosibirsk (1977)
138. Romalo, D. I.: Rev. Roumaine Sci. Techn., Série Electrotechnique et Energétique *13*, 415, (1968)
139. Finkelnburg, W., Maecker, H.: Elektrische Bögen und thermisches Plasma, Handb. Physik, Springer-Verlag, Berlin 1956
140. Nachman, M.: Rev. Htes. Temp. et Réfract. 10, *65*, (1973)
141. Pfender, E.: Electric Arcs and Gas Heaters, in: Gaseous Electronics (ed. M. N. Hirsh, H. J. Oskam), Acad. Press. 1978
142. Hoyaux, M. F.: Arc Physics, (Appl. Phys. Engng. vol. 8), Springer-Verlag, Berlin 1968
143. Bourdin, E.: Calcul approximatif d'un tuyère de torche à plasma par application du modèle à propriétés constantes, Rap. Int. Lab. Thermodynamique, april 1978
144. Schoumaker, H. R. P.: Fours à chauffage plasma, in: Conf. Proc. Internat. Round Table on Study and Application of Transport Phenomena in Thermal Plasma, Odeillo, sept. 1975
145. Geideman, W. A. Vr.: A general discussion of plasma generator design consideration and operating characteristics, Plasmadyne Corporation, PRE 107
146. Schoeck, P. A.: Ph. D. Thesis Univ. Minnesota 1961
147. Joukov, M. F., Koroteel, Ourioukob, B. A.: Hydrodynamique appliquée des plasmas thermiques (Koutatrieladze S. S. ed.) Nauka, Novosibirsk 1975
148. Joukov, M. F., Kourotchkine, Poustogarov, A. V.: Etude des plasmatrons utilisant des gaz plasmagènes soufflés à travers un étage rapporté entre les électrodes, p. 82 in: Investigation expérimentale des plasmatrons (MF Joukov, ed.) Nauka, Novosibirsk 1977
149. Gey, M. G., Kemeny, G. A.: Method of direct ore reduction using a short gar arc heater, U.S. pat. 3, 765, 870, oct. 16 1973
150. Bonet, C. et al.: J. Phys. D., Appl. Phys. *9*, L141, (1976)
151. Shakov, M. F., Smolyakov, V. Ya., Urgukov, B. A.: Electric-Arc Heaters of Gases Nauka, Moscow 1973
152. Sayce, I. G.: Plasma processes in extractive metallurgy, in: Proc. Symp. Adv. in Extractive Metallurgy and Refining, London, oct. 4–6, 1971
153. Hamblyn, S. M. L.: Minerals Sci. Eng. *9*, 151, (1977)
154. Gold, D.: Revue de la technologie des plasmas thermiques, Nouvelles Applications de l'Electricité, Coll. Internat., Toulouse, 7–11 march 1977, Contribution VI-3
155. Fauchais, P., Bourdin, E.: Ceramics and plasma, in: Proc. III Internat. Symp. Plasma Chemistry, Limoges, 13–19 july 1977
156. Bonet, C.: Le four à plasma en minéralurgie et en métallurgie extractive Journée d'Etude de la SEE, J. 27 april 1978
157. Maecker, H.: Z. Physik, *136*, 119 (1953)
158. Pateyron, B.: Interaction entre phénomènes de transport et processus de métallurgie par plasma, Rapport interne Laboratoire de Thermodynamique Univers. Limoges, june 1978
159. Bhattacharya, D., Gauvin, W. H.: J. Appl. Phys. *47*, 4863 (1976)
160. Wilks, P. H. et al.: Chem. Eng. Prog. *68*, 82 (1972)
161. Sheer, C., Korman, S., Kang, S. F.: Investigation of convective arcs for the simulation of re-entry aerodynamic heating, AFOSR-TR-74-1505, Contract F-44 620-69-C-0104 (1974)
162. Bayliss, R. K., Bryant, J. W., Sayce, I. G.: Plasma dissociation of zircon sands, Proc. (S-5-2) III Internat. Symp. Plasma Chemistry, Limoges, 12–19 july 1977

163. Kubanek, G. R., Munz, R. J., Gauvin, W. H.: Plasma decomposition of molybdenum disulphide, Progr. Rep., Proc. (S-5-4) III Internat. Symp. Plasma Chemistry, Limoges, 13–19 july 1977
164. Bonet, C. et al.: Rev. Int. Htes Temp. et Refract. *11*, 11 (1974)
165. Pickles, C. A. et al.: Transactions ISIJ *18*, 369 (1978)
166. Whyman, D.: A rotating wall DC Arc Plasma Furnace, J. Sci. Instr. *44*, 525 (1967), US Pat. 3, 433, 991, march 18 1969
167. Tylko, J. K.: High Temperature treatment of materials, Can. Pat. 957733, Granted to Tetronics Ltd., 1974
168. Tylko, J. K.: The in-flight manufacture of irons and steels in an expanded precessive plasma reactor, Proc. III Internat. Symp. Plasma Chemistry, Limoges, 12–19 july 1977
169. Arc-Coal Dev. Final Rep., May 1966, April 1972, AVCO Corp. Systems Division, Lowel, Massachussetts, USA, 91851
170. Grosse, A. V. et al.: Mater. Res. Stand. *5*, 173 (1963)
171. Foex, M., Delmas, R.: C.R.A.S., Paris *C1*, 9 (1967)
172. Sayce, I. G., Selton, B.: Special Ceramics. Brit. Ceram. Res. Assoc., *5*, 157 (1972)
173. Yerouchalmi, D. et al.: High Temperatures — High Pressures *3*, 271 (1971)
174. Howie, F. H., Sayce, I. G.: Rev. Int. Htes. Temp. et Refract. *11*, 169 (1974)
175. Chase, J. D., Skriven, J. F.: Process for the benefication of titaniferous ores utilising a hot wall continuous plasma reactor, U.S. Pat. 3.856.918, (1974)
176. M. H. D. Research Inc., U.K. Pat. 1, 205, 576 (1970)
177. Gold, R. G., Sandall, W. R., Cheplick, P. G., Mac-Rae, D. R.: Plasma reduction of iron oxide with hydrogen and natural gas at 100 kW and 1 MW, Internat. Round Table on Study and Application of Transport Phenomena in Thermal Plasmas, IUPAC, Odeillo, 12–16 sept. 1975, Proc.
178. Magnolo, G.: Can. Min. Metall. Bull. *57*, 57 (1964)
179. Fiedler, H. et al.: Results of plasma melting of steel, Fith Internat. Symp. Electroslag Remelting and Other Special Melting Processes Pittsburg, 1974
180. Schoumaker, H. R. P.: Fours à plasma, in: Proc. Internat. Round Table on Study and Application of Transport Phenomena in Thermal Plasmas, IUPAC, Odeillo, 12–16 sept. 1975
181. Paton, B. et al.: Plasma arc remelting in a copper water-cooled crystallizer as a new method of improving metal and alloy properties, Proc. Third Internat. Symp. Electroslag Melting Processes, Pittsburgh, USA 1971
182. Kinoshito, T.: Shinka *17*, 61 (1975)
183. Sayce, I. G.: Pure and Appl. Chem. *48*, 215 (1976)
184. Sheer, C., Korman, S.: Arcs in inert Atmospheres and Vacuum, p. 169, Wiley, New York, 1956
185. Besson, J. L., Boch, P.: Plasma Spraying of Ceramics, Internat. Round Table Discussion on Special Ceramics for Electronics and Electrical Engineering, Warsaw, 8–11 oct. 1978
186. Madejski, J.: Int. J. Heat and Mass Transfer *19*, 1009 (1976)
187. Madejski, J.: Bull. Acad. Pol. des Sciences *24*, n° 1 (1976)
188. Zoltowski, P.: Rev. Int. Htes. Temp. et Refract. *6*, 65 (1968)
189. Meyer, Von H.: Ber. dtsch. keram. Ges. *41*, 112 (1964)
190. Grisaffe, S. J., Spitzig, W. A.: NASA. Technical Note D-1705 (1963)
191. Jones, H.: Rep. Progr. Phys. *36*, 1425, (1973)
192. Donovan, M.: Brit. Weld. J. *13*, 490 (1966)
193. Kudinov, V. V., Kitaev, F. I., Tsidulko, A. G.: Parosh, Metall. *152*, 38 (1975)
194. Urbain, C.: Bull. Soc. Fr. Ceram. *4*, 5 (1965)
195. Grisaffe, S. J., Spitzig, W. A.: Trans. A.S.M., *56*, 618 (1963)
196. Sitzer, D. H.: Metal Prog. *86*, 128, (1964)
197. Kitahara, S., Hasui, A.: J. Vac. Sc. Technol. *11*, 747 (1974)
198. Marynowski, C. W., Halden, F. A., Farley, E. P.: Electrochem. Tech. *3*, 109, (1965)
199. Hasui, A., Kitahara, J.: Proc. N.R.I.M. (Tokyo), June 1966, pp. 28–30
200. Ingham, H. S., Shepard, A. D.: Metco Flame Spray Handbook (1965)
201. Mitin, B. S., Nagibin, Yu. A.: Izv. Akad. Nauk, SSSR, Neorg. Mater. *7*, 814, (1971)
202. Guyonnet, J.: Brevet C.N.R.S. 124111 and others: 168044 et 7044678
203. Tucker, R. C.: J. Vac. Sci. Technol. *11*, 725, (1974)
204. Muehlberger, E.: Plasmasadyne Rep. (1974)

205. Chevala, O. B., Orlova, L. M., Ryakich, L. M., Lemesh, A. D.: Porosh Metallurgiya *6*, 37, (1976)
206. F. Cabannes, private communication
207. Okada, M., Maruo, H.: Brit. Weld. J. *15*, 371, (1968)
208. McPherson, R.: J. Mats. Sci. *8*, 851, (1973)
209. Sokolova, T. V. et al.: Izv. Akad. Nauk. SSSR, Neorg. Mat. *9*, 611, (1973)
210. Eichorn, F., Metzler, J., Eysel, W.: Metalloberfläche *26*, 212 (1972)
211. Huffachine, J. B., Thomas, A. G.: Powder. Met. *7*, 290, (1964)
212. Thompson, V. S., Wittemore, O. J.: Ceram. Bull. *47*, 637, (1968)
213. Guyonnet, J., Borie, A., Fauchais, P.: Projection au chalumeau à plasma de revêtements céramiques à coefficients de frottement variables, Proc. Internat. Round Table on Study and Applications of Transport Phenomena in Thermal Plasmas, Odeillo, 12–16 sept. 1975
214. Buckley, D. H. et al.: Wear, *11*, 405 (1968)
215. Stejin, R. P.: J. Appl. Phys. *32*, 1951, (1961)
216. Mock, J. A.: Mat. Eng. *4*, 101 (1974)
217. Moss, M., Cyrus, W. L., Schuster, D. M.: Ceram. Bull. *51*, 167, (1972)
218. Manual: Plasma Processes in Metallurgy and Process Engineering of Nonorganic Materials, Nauka, Moscow 1973
219. Hamblyn, S. M. L.: A review of application of plasma technology with particular reference to ferro-alloy production, Nat. Inst. Metallurgy, Rep. n° 1895, Randburg, South Africa, 14 april 1977
220. Waldie, B.: Chem. Eng. [Lond.] *262*, 188 (1972)
221. Rutkowski, W., Szteke, W.: Possible uses of plasma in powder metallurgy, with particular reference to spheroïdisation, Rep. 938/XIV/R (1968), Inst. Nuclear Res. Warsaw
222. Anon, : Chem. Eng. *24*, 56 (1975)
223. Bonnet, C., Vallbona, G., Foex, M.: Bull. Soc. Franc. Ceramique *94*, 27 (1972)
224. Gani, M. S. J., McPherson, R.: J. Aust. Ceram. Soc.
225. Beal, G. H.: High Temperature Oxides Part IV Refractory Glasses, Glass Ceramics and Ceramics (ed. A. M. Alper), Academic Press London 1971
226. Waldie, B.: S. Mater. Sci. *4*, 648 (1969)
227. Chales, J. A. et al.: Trans. Inst. Min. Metall. *79*, 154 (1970)
228. Gilman, W. S., Seabough, P. W., Sullenger, D. B.: Science *160*, 1239 (1968)
229. Hedger, H. J., Hall, A. R.: Powder Metall. *8*, 65 (1961)
230. Fey, M. G., Wolf, C. B., Harvey, J. F.: Magnetite spheroïdization using an A.C. arc heater, Proc. Internat. Round Table: Transport Phenomena in Thermal Plasmas, Odeillo, (France), 12–16th sept 1975
231. Evans, A. M., Williamson, J. P. H.: J. Mats. Sc. *12*, 779 (1977)
232. Evans, A. M., Williamson, J. P. H.: ibid. *14*, 680 (1979)
233. Garvie, R.: ibid. *14*, 817 (1979)
234. Barret, M. F., Howie, F. H., Sauce, I. G.: Trans. Inst. Min. Metal. (Sec C.) *84*, 231 (1975)
235. Borgianni, C. et al.: Combustion and Flame *13*, 181 (1969)
236. Johnston, P. D.: ibid. *18*, 373 (1972)
237. Bonet, C., Daguenet, M., Dumargue, P.: Int. J. Heat. Mass Transfer *17*, 643 (1974)
238. Capitelli, M. et al.: Combust. Flame *15*, 23 (1970)
239. Barnes, W. R.: U.K. Pat. 1.211.702 (1970)
240. Korman, S. Sheer, C.: Electrochem. Soc. Symp., Philadelphia 1966
241. Bonet, C., Vallbona, G., Foex, M.: Bull. Soc. Fr. Ceram. *94*, 27 (1972)
242. Korman, S. et al.: Fine particles (ed. W. E. Kuhn), p. 153, Electrochem. Soc., Princeton 1974
243. Everest, D. A., Sayce, I. G., Selton, B.: J. Mater Sci. *6*, 218 (1971)
244. Barret, M. F. et al.: Faraday Symp. *8*, 149 (1973)
245. Barret, M. F., Howie, F. H., Sayce, I. G.: Trans. Inst. Min. Metall. *84*, C231 (1975)
246. Arkless, K. Ceaver, D.: Brit. Pat. 1226082 (1966)
247. De Pous, O.: Proc. 3rd Intl. Symp. Plasma Chem. IUPAC, Université de Limoges, (ed. P. Fauchais), paper S. 4.7., july 1977
248. Audsley, A., Bayliss, R. K.: J. Appl. Chem. *19*, 33 (1969)
249. Ulrich, G.: Combust. Sci. and Technol. *4*, 47 (1971)
250. Cozzi, C., Cadoran, D.: Combust. Sc. and Tech. *5*, 213 (1972)

251. Gani, M. S. J., McPherson, R.: J. Mats, Sc. *12*, 999 (1977)
252. Gani, M. S. J., McPherson, R.: "Structure of Al_2O_3—TiO_2 submicron particles", in: VIIIth Int. Congr. Electron Microscopy, Camberra 1974, vol. 1
253. Mackenzie, J. K. Shuttleworth, R.: Proc. Phys. Soc. *B62*, (1949), 833
254. McPherson, R.: Proc. Intl. Round Table Study Application of Transport Phenomena in Thermal Plasmas (ed. C. Bonnet) IUPAC/CNRS, Odeillo 1975
255. Holmgren, J. D., Gibson, J. O., Sheer, C.: Ultrafine Particles (ed. W. E. Kahn), Wiley, New York (1963)
256. McPherson, R.: J. Mats. Sc. *8*, 859 (1973)
257. Dong Hi Lee, McPherson, R.: J. Mats Sc., in press.
258. McPherson, R.: J. Mat. Sc. *15*, 3141 (1980)
259. Dong Hi Lee, McPherson, R.: Proc. 3rd Intl. Symp. Plasma Chem., IUPAC/Univers. Limoges. (ed. P. Fauchais), paper S. 4.10, july 1977
260. Dong Hi Lee, McPherson, R.: Proc. 9th Aust. Ceramics, Conf. Melbourne 1978
261. Stokes, C. S., Knipe, W. W.: Ind. Eng. Chem. *52*, 287, (1960)
262. Stokes, C. S., Cahill, J. A., Correa, J. J., Grosse, A. V.: Plasma jet chemistry a final repport, Air Force office of Sci. Res., Grant 62–196, Res. Inst. Temple Univers., Philadelphia, Pa., dec. 1964
263. Stokes, C. S.: Chem. Engng. *72*, 191, (1965)
264. DeVynck, I. A., Bosch, F. M.: Silicates industriels (1970–1971)
265. Matsumoto, O.: J. Elec. Soc. Japan *36*, 153, (1968)
266. Matsumoto, O., Ohtao, K.: Denki Kagaku: *41*, 17, (1973)
267. Akashi, K., Yoshido, T., Nakagawa, K.: 3rd Intl. Symp. Plasma Chem. IUPAC/Université de Limoges (ed. P. Fauchais), paper S. 3.2., july 1977
268. Hayashi, Y., Okada, K., Koide, S.: J. Ceram. Assoc. Japan *76*, 307, (1968)
269. Opfermann, W.: Monatsber. Deutsch. Akad. Wiss. Berlin *2*, 92, (1964)
270. Winterhager, H. Harnusch, K.: Ber. deut. keram. Ges. *4*, 181, (1969)
271. Harnusch, K., Winterhager, H.: Metallwiss. Technik *25*, 1061, (1970)
272. Matsumoto, O.: J. Electrochem. Soc. Japan *36*, 213 (1968)
273. Marsumoto, O., Kawai, Y.: Denki Kagaku *40*, 271, (1972)
274. Matsumoto, O., Abe, K.: ibid. *43*, 75, (1975)
275. Matsumoto, O., Tsuda, T., Makumoki, A., Hayakawa, Y.: ibid. *41*, 352, (1972)
276. Leprince-Ringuet, F., Lejus, A. M., Collongue, R.: C.R. Acad. Sc. Paris, *258*, G8, 221, (1964)
277. Matsumoto, O., Hayakawa, Y.: J. Electrochem. Soc. Japan *34*, 216, (1966)
278. Neuenschwander, E.: J. Less. Common Metals *11*, 365, (1966)
279. Matsumoto, O.: J. Electrochem. Soc. Japan *36*, 207 (1968)
280. Matsumoto, O., Shirato, Y., Migazaki, M.: J. Electrochem. Soc. Japan, *36*, 219 (1968).
281. Matsumoto, O., Shirato, Y.: Denki Kagaku *39*, 82, (1971)
282. Becherescu, D.: Stud. Cerc. Chim. Roman. *16*, 879, (1968)
283. Szymanski, A., Huczko, A., Podgorski, A.: Z. Phys. Chem. *259*, 5289, (1978)
284. Szymanski, A., Podgorski, A., Huczko, A.: 3rd Intl. Symp. Plasma Chem. IUPAC/Univ. Limoges (ed. P. Fauchais) paper S. 4.13, july 1977
285. Canteloup, J., Mocelin, A.: Special Ceramics 6 (ed. Papper), 209, (1975)
286. Matsumoto, O., Hayakawa, Y.: J. Electroc. Soc. Japan *36*, 146 (1968)
287. Bosch, F. M., de Wynck, I. A.: Silicate Ind. *27*, 587, (1962)
288. Matsumoto, O., Shirato, Y., Hayakawa, Y.: J. Electrochem. Soc. Japan *37*, 151, (1969)
289. Exell, S. F. et al.: Fine Particles, 2nd Int. Conf. Electrochem. Soc. 165, (1974)
290. Matsumoto, O., Shirato, Y., Hayakawa, Y.: J. Electrochem. Soc. Japan *37*, 151 (1969)
291. Matsumoto, O., Miyazoki, T.: High Temp. Sci. *5*, 40, (1973)
292. Matsumoto, O., Saito, M.: Bul. Chem. Soc. Japan *48*, 351, (1975)
293. Neuenswander, E., Schnett, K., Scheller, W.: Finely dispersed Carbides and process for their production. US Pat. 3,340,020, 5 Sept. 1967
294. Swaney, L. R.: Pat. 2.018.962 (26 june 1970) (France)
295. de Pous, O., Mollard, F., Lux, B.: Proc. 3rd Intl. Symp. Plasma Chem. IUPAC/Univ. Limoges (ed. P. Fauchais), paper 5.4.7. july (1977)
296. Salinger, R. M.: Ind. Eng. Chem., Prod. Res. Develop. *11*, 230, (1972)

297. Matsumoto, O., Saito, M.: High Temp. Science *6*, 135 (1974)
298. Matsumoto, O., Mitarai, Y.: Denki Kagaku *44*, 815 (1976)
299. Matsumoto, O., Saito, M., Yoshino, T.: Bull. Chem. Soc. Japan *47*, 2226, (1974)
300. Matsumoto, O., Yaguchi, Y.: Proc. 3rd Intl. Symp. Plasma Chem. IUPAC/Univ. Limoges (ed. P. Fauchais), paper 5.4.2. july 1977
301. Wickens, A. J.: Chem. Ind. 3 april 1976
302. MacKinnon, I. M., Reuben, B. G.: J. Electrochem. Soc. *122*, 806 (1975)
303. MacKinnon, I. M., Hamblyn, S. M. L., Reuben, B. G.: 3rd Intl. Symp. Plasma Chem. IUPAC/Unive. Limoges (ed. P. Fauchais), paper 5.4.3., july 1977
304. Miller, R. C., Ayen, R. J.: I.C.E.C. Proc. Design. Dew *8*, 370, (1969)
305. Yakashi, K., Ohno, S., Ishizuka R., Yoshida, T.: 3rd Intl. Symp. on Plasma Chem. IUPAC/Univ. Limoges (ed. P. Fauchais) paper 5.4.6., july 1977
306. Diana, M., Luponio, C., Rosso, G.: Rev. Phys. appliquée, *12*, 1237, (1977)
307. Hamblyn, S. M. L., Reuben, B. G., Thompson, R.: Special Ceramics 5 (Ed. P. Popper) Brit. Ceram. Res. Assoc. Stoke on Trend 1972

FSC
www.fsc.org
MIX
Papier | Fördert gute Waldnutzung
FSC® C083411

Zeitfracht Medien GmbH
Ferdinand-Jühlke-Straße 7
99095 Erfurt, Deutschland
produktsicherheit@kolibri360.de